FOLKLORE OF FEBRUARY

A SEASONAL FOLKLORE GUIDE TO NATURE TRADITIONS,
PLANT LORE AND WINTER'S TURNING

MIRA LINDEN

CONTENTS

At the Edge of Winter ... v
February as Threshold Time ... vii

PART 1
FEBRUARY AROUND THE WORLD: TRADITIONS OF RENEWAL

1. Brigid's Day and the Quickening of Light (Ireland and Scotland) ... 3
2. Candlemas and the Hearth of Winter (Europe) ... 9
3. Setsubun and the Sweeping Out of Winter (Japan) ... 15
4. Lunar New Year: Blossoms of Good Fortune (China and Diaspora Communities When It Falls in February) ... 21
5. Groundhog Day and Animal Weather Wisdom (North America) ... 27
6. Carnival: Winter's Mask and Mischief (Global traditions) ... 33
7. Love, Lots and Divination: Midwinter Courtship Beliefs (Europe and elsewhere) ... 39

PART 2
PLANTS, TREES AND MATERIALS OF FEBRUARY

8. Snowdrops ... 47
   Lanterns at Ground level
9. Hazel ... 52
   Wisdom at the Woodland Edge
10. Willow ... 57
    Flexibility, Flow and Early Growth
11. Evergreen Guardians ... 62
    Pine, Yew and Juniper
12. Milk ... 67
    and the Quickening Herd
13. Wax and Flame ... 72
    The Materials of Returning Light
14. Ash and Ember ... 77
    The Memory of the Winter Hearth

PART 3
SKY, EARTH AND THE SCIENCE BEHIND THE STORIES

15. Why February Is a Turning Point — 85
16. The Hunger Moon — 90
    Old Names for Harsh Days
17. Reading the Sky — 95
    Reading the Sky
18. Songs in the Cold — 100
    Songs in the Cold
19. Stirrings Beneath the Soil — 106

PART 4
LIVING THE MONTH: PRACTICES FOR MODERN READERS

20. Simple Rituals of Light and Purification — 113
21. Noticing the First Signs of Spring — 120
22. Working with February Plants and Materials — 127
23. A Personal February — 132
    Reflection Pages

PART 5
APPENDICES

24. February Moon Phases and Sky Notes — 139
25. A Global List of Festivals and Plant Markers — 142
26. Glossary of Botanical and Folkloric Terms — 146

AT THE EDGE OF WINTER

There is a particular stillness that settles over the land in February, a quiet that is neither the deep rest of midwinter nor the lively stirring of spring. It is a pause that feels almost fragile, as if the world is holding its breath. Light returns in small but noticeable steps. Birds begin to test their voices. Shoots push through frost with a confidence that seems almost impossible. Everything waits, yet nothing is truly still.

Across many cultures, February has long been understood as a time of thresholds.

People felt the subtle change in the air and shaped stories to make sense of it. Fires were carried from one season to the next. Houses were swept clean. Milk began to flow as new life entered barns and fields. Festivals of purification, protection, renewal and hope appeared in every corner of the world. Even the smallest signs in nature were given meaning. A snowdrop was not only a flower but a message. A shadow on a bright morning could be a sign of what the coming weeks might hold.

Science explains many of the changes we see in this month. The growing length of daylight shifts plant chemistry and animal behavior. Soil microbes wake in the cold ground. Sap begins its slow rise. Birds answer invisible cues that prepare them for pairing. Yet these facts do not diminish the older stories. If anything, the blend of myth and mechanism reveals a deeper truth. Humans have always sensed the rhythm of the seasons, long before we could measure it. Intuition guided traditions that remain surprisingly accurate even today.

This book is an invitation to step gently into that meeting place where folklore and natural history shape our understanding of the year.

February often passes unnoticed, overshadowed by the darker days that came before and the bright promise of spring that follows. Yet this month offers something rare. It allows us to witness change in its earliest form, when it is still delicate and easily overlooked.

To read the folklore of February is to see how people across generations paid attention to the world. They noticed the slight tipping of light across a hearthstone. They watched the flight of crows, the blush of plum blossoms, the first swelling of buds. They honored these signs with rituals that brought communities together and offered meaning to ordinary days. Their stories are small bridges between what they observed and what they hoped for.

As you move through these pages, you may find that February begins to open itself to you in new ways.

The plants, traditions and quiet sciences of the season become companions. You may notice how snow sounds underfoot or how early birds seem to sing against the cold rather than with it. You may feel the subtle rise of energy in the natural world, a reminder that life returns long before spring fully arrives.

At the edge of winter, we stand with centuries of storytellers, farmers, healers and observers who felt the first tremble of renewal in this month. Their voices echo through these traditions. Their insights live in the plants that bloom, the customs that endure and the wisdom carried in the land itself.

May this book help you meet February with curiosity, gentleness and a renewed sense of connection to the turning world.

FEBRUARY AS THRESHOLD TIME

February is a month that many people overlook. It arrives in the quiet after winter holidays and seems to drift by in a soft haze of cold mornings and early sunsets. Yet beneath this gentleness lies a powerful truth. February is a turning point in the natural year, a moment when winter begins to loosen its grip and the first signs of renewal appear. These changes are often small and easily missed. They might be a bird call heard through a window, a swell of bud on a seemingly bare branch or a faint shift in the quality of light. Still, they carry the promise of everything that will come.

> *For countless generations, people paid close attention to this month. They understood how important it was to watch for subtle shifts in the world around them.*

What they observed became stories, customs and rituals that helped them make sense of the slow movement from darkness to light. February became a time of cleansing, renewal, protection and preparation. Fire festivals marked the growing strength of the sun. Offerings were made to hearth and home. Animals were studied for signs of the weather ahead. Early blooming plants were treated with reverence and used to predict the fortunes of the season.

These traditions may seem distant to modern readers, yet they reveal something timeless. Humans have always sought meaning in the natural world. A flower breaking through snow offered hope. A lengthening day signaled safety. A returning bird proved that the cold season would not last forever.

> *Even when scientific understanding was limited, people sensed the patterns of the earth and shaped their beliefs around them.*

Today we know far more about the mechanisms behind these changes. Light levels begin to rise shortly after the winter solstice. Plants prepare for spring long before their leaves appear. Animals respond to hormonal signals triggered by day length rather than temperature alone. Soil communities wake from dormancy while the ground is still cold. Each of these processes contributes to the quiet transformation that defines February.

Yet scientific knowledge does not replace the value of folklore. Instead, the two enrich each other. Folklore preserves the emotional, cultural and symbolic responses of earlier communities. Science explains the physical processes that inspired those responses. Together they offer a full picture of how humans relate to the seasons. That is the heart of this book. It is a meeting place where cultural stories and natural history are held with equal respect.

> *The chapters that follow explore how February is experienced across different regions and traditions.*

You will encounter fire rituals, purification customs, seasonal foods and protective charms, each shaped by unique landscapes and local histories. You will also discover the plants, trees and natural materials that define this month. Their cycles reveal why certain beliefs formed and why some plants became powerful symbols in seasonal lore.

This book is not meant to instruct a single way of observing February. Instead, it encourages you to notice what this month already offers. You might begin to see small details in the world that once seemed ordinary. The rhythm of bird calls on a cold morning. The soft gleam of wax in candlelight. The scent of evergreen branches. The promise held inside a tightly closed bud.

> *February teaches us that change begins quietly. It does not rush or demand attention. It works through small steps, hidden growth and gradual awakening.*

It invites patience and presence. By exploring the folklore and natural history of this month, we learn to appreciate the subtle ways the world prepares for renewal.

May this introduction open the door to a season often forgotten but rich with meaning. May it encourage you to look closely, listen gently and rediscover the enchantment of winter's turning.

PART 1
FEBRUARY AROUND THE WORLD:
TRADITIONS OF RENEWAL

1 / BRIGID'S DAY AND THE QUICKENING OF LIGHT (IRELAND AND SCOTLAND)

FOR THE CELTIC WORLD, the first days of February mark a gentle but profound shift in the year. Winter is still present. Frost tips the grass, winds sweep across open fields and the memory of long nights still lingers. Yet something unmistakable changes. Light stretches a little further across the sky each afternoon. Dawn arrives with a faint warmth rather than complete darkness. People of early Ireland and Scotland felt this change deeply. They honored it through customs that welcomed the returning sun and prepared their communities for the new season ahead.

. . .

Brigid's Day falls on the first of February and sits at the midpoint between the winter solstice and the spring equinox. Known as Imbolc in older traditions, it was a festival that celebrated the promise of renewal long before spring flowers filled the landscape. It honored Brigid, a figure whose roles blended the sacred and the everyday. She was associated with fire, healing, poetry and the tending of animals. In a world where survival depended on careful attention to the land, Brigid embodied the hope and strength people needed to move from darkness toward the growing light.

Fire Returning

At the heart of Brigid's Day lies the theme of returning fire. Winter fires were once essential for life. They warmed homes, dried clothing and provided the only sources of light. As daylight increased in early February, people recognized a symbolic parallel between the sun in the sky and the flames in their hearths. They saw this rising brightness as a sign that nature was awakening.

Families often tended their hearth carefully on the night before Brigid's Day. Some allowed the fire to burn low, believing that Brigid herself might bless the embers as she passed over the land. Others placed a small white cloth or ribbon near the hearth or outside the door, hoping Brigid would touch it as she journeyed. By morning, this cloth was believed to hold healing energy and protection.

Lighting fresh candles or allowing an older flame to flare in the hearth symbolized the strengthening sun. The practice reflected more than simple survival. It carried a quiet spiritual message. Fire was renewal, purity, inspiration and transformation. People greeted the new light with gratitude and a readiness to shift their minds from endurance to preparation.

Milk Lore and Early Lambing

In agricultural communities, February also marked the beginning of lambing season. Ewes carried their young through the darkest days of winter and began to give birth as daylight increased. Milk returned to households at a time when food stores were running low, and this natural cycle became the origin of many customs.

The word Imbolc is often linked with a phrase meaning in the belly, a reference to both pregnancy and the swelling of life hidden under winter frost.

Early milk was precious. It nourished families, provided warmth and supported new livestock. People believed that Brigid watched over animals, especially sheep, and guided this part of the cycle.

Milk was used in blessings, poured in small amounts on thresholds or mixed into celebratory foods. In some households, families set out a small dish of milk by the door as an offering to ensure good health for the flock.

This attention to animals reflects a truth often forgotten in modern times. The survival of an entire community depended on these rhythms. The return of milk meant the land was shifting toward abundance once more

Rush Weaving Traditions

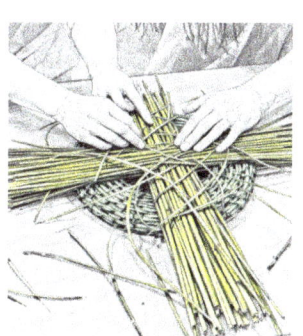

Perhaps the most recognizable tradition of Brigid's Day is the weaving of the Brigid's cross. These simple but beautiful designs are crafted from fresh rushes gathered near streams and wetlands. The cross is not a weapon or emblem of conquest. It is a symbol of protection, peace and household blessing.

Families would sit together on the evening before Brigid's Day and weave crosses while telling stories. The act was tactile and grounding. It connected hands and hearts. Once woven, the crosses were hung above doorways, in stables or near hearths to invite safety for the year ahead.

> **Plant Facts: Rushes**
> Common Rush (Juncus effusus)
>
> • Thrives in wet soils and marshy ground
> • Bends easily without breaking, symbolizing resilience
> • Used historically for floor coverings, baskets and lamp wicks
> • Represents protection and renewal in Irish tradition
>
> Why Rushes for Brigid's Crosses? Rushes were accessible, flexible and tied to water, an element of cleansing and healing. Their green color also symbolized future growth.

. . .

The ritual also linked people directly to the land. Rushes are humble plants, often overlooked. Yet their presence shows where freshwater moves beneath the surface.

By choosing these plants for their sacred symbol, people honored simple materials that carried practical value.

Rushes were used in baskets, bedding, floors and lamps. They were part of everyday life. Turning such materials into symbols of blessing reflected the belief that protection and magic often reside in ordinary things.

Plant Focus : Rushes and White Flowers

Rushes were not the only plants associated with Brigid's Day. White flowers became quiet emblems of purity and new beginnings. In Ireland and Scotland, snowdrops often bloom at this time. Their slim stems and nodding white bells push through frozen soil with surprising confidence. Snowdrops were sometimes called Brigid's blooms or February's fair maids. People believed they carried a gentle power to guard against misfortune and sickness.

. . .

These flowers were not placed on graves or near ill omens as some blossoms were. Instead, they were welcomed into homes and placed near windows to mark the arrival of hope. In a season where color is scarce, a small white flower felt like a message from the earth itself.

Rushes and snowdrops share a similar symbolic language. Both thrive in cold, both bend without breaking and both herald the early moments of change. Their presence reminded communities that life persisted even under winter's weight.

Natural History : Why Light Length Matters Now

Long before scientific tools measured daylight, people recognized that February carried a shift in energy. Today we understand why. During this month, the Northern Hemisphere experiences a discernible increase in light each day. Even slight changes in day length influence plants and animals far more than temperature does. Plants begin preparing for growth weeks before they appear to wake. Increased light affects the chemistry of buds, which swell in response. Snowdrops produce their own heat to melt small pockets of snow around them. Willow and hazel catkins begin to loosen. Birds respond to the growing light with hormonal changes that trigger early courtship behaviors. Livestock are also influenced by light. The timing of lambing is tied to seasonal daylight patterns that have guided sheep for thousands of years.

Plant Facts: Snowdrops
Galanthus nivalis

- One of the earliest blooming flowers in Europe
- Produces heat through a process called thermogenesis
- Often pushes upward through frozen soil
- Associated with purity, renewal and gentle protection

Folklore Note
Snowdrops were sometimes called February's fair maids, a nod to their clean white petals and graceful form.

In this way, folklore is rooted in real observation. People sensed these changes through their daily lives. They knew when animals behaved differently. They felt the shift in morning light. They recognized that fire burned brighter in a home filled with renewed hope.

Brigid's Day is a reminder that the natural world begins its transformation long before spring is visible.

February teaches patience, attentiveness and trust. It invites us to celebrate small changes and to believe in the quiet promise of returning warmth.

Reflection and Ritual for Brigid's Day

Set aside a quiet moment during the first days of February. Light a small candle and notice the quality of the flame. Allow it to represent the returning light of the season. If you have access to rushes, straw or even simple paper strips, weave a small cross or knot pattern. Hold it gently and think about the year ahead. Ask yourself what needs protection, what needs renewal and what needs space to grow.

Place your woven piece near your door, on a shelf or by a window. Let it serve as a reminder that change often begins quietly. The season is shifting. The world is waking. You are part of that unfolding rhythm.

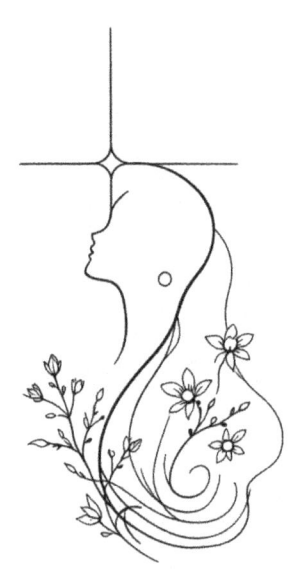

2 / CANDLEMAS AND THE HEARTH OF WINTER (EUROPE)

CANDLEMAS ARRIVES on the second of February and has long been a marker of winter's slow retreat. Though the cold lingers and mornings still bite with frost, people throughout Europe recognized this date as a hinge in the seasonal year. In Christian tradition it honors the purification of Mary and the presentation of the infant Jesus at the temple. Yet beneath the religious layers lie older customs shaped by the land, the sun and the deep human need to measure time through light.

Across villages and towns, Candlemas became a celebration of brightness breaking through darkness. Candles were blessed in churches, carried through homes and placed in windows to honor the lengthening days. For families who lived close to the land, Candlemas was both spiritual and practical. It was a day to assess food stores, tend to tools, ready seeds and calculate how many weeks remained until the first true signs of spring.

Where Imbolc honors the early spark of renewal, Candlemas carries the theme of illumination, both literal and symbolic. It is a day that asks people to notice not only the returning sun but also the warmth and guidance that fire offers through the remainder of winter.

Fire and the Hearth of Midwinter

For many European households, the hearth was the center of life. It provided warmth, cooked food and offered light during long nights. Candlemas traditions reflect this central relationship. Before the day arrived, homes were often swept clean. Ashes were removed from the hearth and replaced with fresh logs. The act of cleaning symbolized clearing away the weight of winter and preparing for the next stage of the year.

Beeswax candles played an important role in these customs.

Their clean scent and steady flame were associated with purity and blessing.

Families carried newly blessed candles through each room of the house to protect against misfortune and bless the household. The glow of beeswax was believed to reveal truth, dispel fear and strengthen the resilience of the occupants.

The hearth also served as a reminder of community. In many regions people visited neighbors, exchanged candles or offered small tokens of hospitality. A bright flame signaled welcome, warmth and shared hope for the coming season.

if Candlemas be fair and bright, winter has another bite... if Candlemas brings cloud and rain, winter will not come again

This well known rhyme reflects older beliefs about weather prediction. A bright sky at Candlemas meant winter was gathering strength for a final effort. A dark or rainy day promised an early spring.

Weather Lore and Seasonal Reading

The practice of reading weather signs at Candlemas is widespread across Europe. In some regions people watched the behavior of animals, especially small hibernating creatures. The appearance of a hedgehog was considered a sign of coming weather. If the creature emerged and saw a clear shadow it retreated again, warning that winter had not finished its work. This tradition later influenced the Groundhog Day custom that developed in North America.

Weather at Candlemas mattered because it affected planting decisions. Farmers needed to know how long frost would hold. A bright day often meant colder weeks ahead because clear skies allowed nighttime temperatures to drop. Cloudy weather suggested a milder stretch of days, which helped soil begin its slow thaw.

While modern forecasting relies on atmospheric data rather than folklore, there is a kernel of truth hidden in these sayings. Climate patterns in late winter often follow predictable rhythms that people learned to read through long observation.

Candles, Wax and Symbolic Light

Candles were more than sources of light. They held symbolic power. Beeswax, in particular, carried spiritual and practical significance. It burned cleanly, produced a warm glow and filled the room with a gentle natural aroma. Wax represented transformation, patience and the slow work of the natural world. Just as bees turned pollen into light, people believed that spiritual strength could grow quietly within the home.

Many traditions included crafting or gathering candles specifically for Candlemas. Some families saved small pieces of wax to use throughout the year in times of difficulty or change.

The candle blessed at Candlemas was often the one lit during storms, illness or moments when guidance was needed.

The purity of wax also connected Candlemas to themes of cleansing and renewal. A fresh candle symbolized a fresh start. Lighting one signaled readiness for whatever the lengthening days might bring.

> **Material Facts: Beeswax**
>
> - Produced by honeybees to build comb structures
> - Burns longer and cleaner than tallow candles
> - Carries natural antibacterial and air purifying qualities
> - Symbol of purity in many European traditions
>
> **Historical Note**
> Beeswax candles were once valuable gifts. They were used for important rituals and celebrations because of their longevity and bright, steady flame.

Plant Symbols and the First Light of Growth

Although Candlemas is best known for its candle rituals, it is also tied to the natural world. In many regions people watched for snowdrops, early crocuses and other pale flowers that followed February's light.

These blooms were treated as gentle heralds of spring. Their appearance was considered a blessing, a sign that the earth was beginning to breathe again. Evergreens also played roles in Candlemas customs. Branches of holly, ivy or juniper were sometimes burned to cleanse the home and honor the transition of the season.

> *Plants symbolized endurance through winter, while white flowers symbolized the first opening toward spring.*

Light in the Natural World

The growing length of daylight in early February influences more than human traditions. Plants shift their internal processes in response to increased light. Buds begin their slow preparation for blooming. Stored sugars in roots and bulbs fuel early growth. Animals adjust their behavior to match new cues. Birds practice tentative songs. Foxes and other mammals begin their mating cycles. Soil microbes wake under the frost and begin nutrient exchanges that will support spring growth.

Even though temperatures may stay cold, life responds instinctively to the sun. Candlemas is an acknowledgment of this truth. It is a celebration of small changes that carry great significance.

Plant Facts: Early February Flowers

Snowdrops (Galanthus nivalis)
• Often the earliest to bloom
• Associated with purity and renewal
• Thrive through cold and frost

Crocuses (Crocus vernus)
• Emerge soon after snowdrops in many regions
• Bright colors symbolize hope and awakening
• Respond strongly to shifts in light and soil warmth

Reflection and Ritual for Candlemas

Take a moment on or near the second of February to honor the growing light. Clear a small space in your home. Wipe dust from a shelf or table and place a candle at its center. Light the flame and notice the soft glow it casts. Let it remind you that the season is shifting one quiet day at a time.
Consider what you wish to illuminate in your own life. Think about what needs clarity, warmth or renewed attention. Write a single word or intention on a small piece of paper and place it beneath the candle holder or in a book you cherish.

If you have access to snowdrops or another early flower, place a single bloom nearby. Let the candle and flower stand together as symbols of endurance and awakening. Allow the moment to be simple, peaceful and full of promise.

3 / SETSUBUN AND THE SWEEPING OUT OF WINTER (JAPAN)

ACROSS JAPAN, the arrival of Setsubun marks the turning of the seasonal year. Although frost may still whiten the fields and winds may blow sharp from the north, Setsubun signals the beginning of a new phase. It comes just before Risshun, the solar term known as the first day of spring. On the surface it may feel too early to think of warm days or blossoms, but for centuries this festival has represented the moment when people symbolically sweep winter away and open the door to good fortune.

. . .

Setsubun traditions blend practicality, community spirit and a touch of playful mischief. Families cleanse their homes through ritual actions, cast out unwanted influences and invite new energy inside. Central to the celebration are roasted soybeans, which serve as tools of renewal, protection and joy. Through these small, humble seeds people reinforce a connection to seasonal change and express gratitude for the cycle of life preparing to unfold.

Bean Casting and Doorway Guards

The most iconic part of Setsubun is mamemaki, the casting of beans. Families gather near doorways or porches, bowls of roasted soybeans in hand, ready to scatter them with purpose and laughter. The phrase Oni wa soto, fuku wa uchi is spoken aloud. Evil out, fortune in. With each toss of beans, negative influences are symbolically sent away and good luck is invited into the home.

Roasted soybeans are chosen intentionally. Raw beans can sprout, symbolizing potential for growth, but roasted beans are safe to scatter without risk of unwanted plants. This detail connects the custom to agricultural wisdom. Communities knew how valuable seeds were. To use them in ritual required care and respect.

Some regions place small bamboo baskets at doorways to catch beans thrown outward. Others scatter beans inside rooms to purify interior spaces. Children often take great joy in the process, which brings laughter to a ritual that is both meaningful and lighthearted.

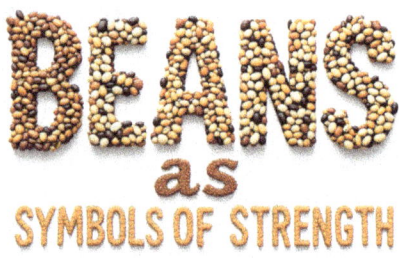

In Japanese folk belief, beans hold life giving power. Their ability to nourish the body and grow into sturdy plants grants them symbolic strength. Casting beans is an act of sending protective energy into the world.

Oni, Boundaries and Purification

Oni, often described as demons or troublesome spirits, play a central role in Setsubun. These beings are not pure evil. They represent chaos, misfortune or lingering energy that disrupts harmony. In folklore oni appear during moments of transition, testing the boundaries between seasons.

Setsubun takes these stories and transforms them into a playful, cathartic event.

People wear oni masks or encourage someone in the family to embody the oni for a few minutes, running through the house or standing at the door while others throw beans at them. The scene is filled with laughter, yet it carries an important message. By facing and casting out the oni, families symbolically remove fear, illness and hardship from their lives.

Doorways hold special significance during Setsubun.

They are natural boundaries between inside and outside, safety and uncertainty, winter and spring. Placing roasted beans, woven ornaments or protective charms near doorways strengthens these thresholds and helps define where comfort begins.

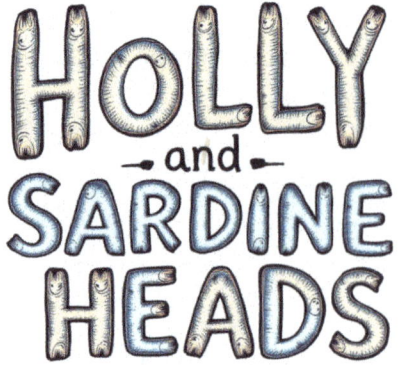

In some regions, people hang a holly branch with a grilled sardine head at the entrance. Holly leaves with their sharp edges repel unwanted spirits, while the sardine scent was believed to deter harmful influences. This talisman is known as yakuyoke, a charm to block misfortune.

Plant Focus - Roasted Soybeans and Holly Leaves

Roasted soybeans, called fukumame or fortune beans, are the heart of Setsubun. They represent vitality, nourishment and the power to influence seasonal luck.

Their golden brown surface evokes warmth in a cold month. After throwing beans, it is customary to eat the number of beans corresponding to one's age plus one more for the coming year.

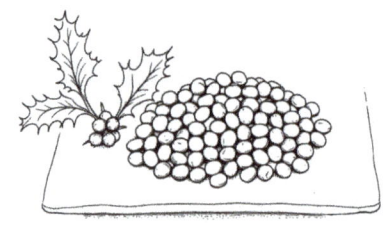

This act is believed to strengthen health and protect the individual throughout the cycle ahead.

Holly leaves, with their sharp points and glossy surfaces, appear often in protective rituals.

Their structure resembles natural armor.

In winter when many plants are bare, holly retains its strength and color, becoming a symbol of endurance.

The pairing of beans and holly creates a balanced charm, one representing inner vitality and the other outer protection.

. . .

Science and the Seasonal Calendar

What "Spring Begins" Means

Although Setsubun occurs during winter weather, it aligns with the traditional Japanese calendar of solar terms. These terms divide the year into twenty four segments based on the movement of the sun rather than the temperature of the air. Risshun, the first day of spring, begins immediately after Setsubun. This does not claim warm days have arrived. Instead it acknowledges the shift toward increasing light and the subtle biological changes already underway.

> **Plant Facts and Material Facts:**
>
> **Roasted Soybeans**
> • Rich in protein and life sustaining nutrients
> • Chosen for their association with growth and resilience
> • Roasting prevents germination, making them safe to scatter
> • Eaten for health, luck and longevity during Setsubun
>
> **Holly**
> (Ilex aquifolium in European varieties, Ilex serrata in Japan)
> • Evergreen shrub known for protective qualities
> • Sharp leaves used to guard entrances
> • Symbol of strength through winter
> • Often paired with strong scents for purifying charms

As daylight grows longer, plants in Japan begin their slow awakening. Buds absorb more light and begin to swell. Sap moves gently within trees. Birds adjust their calls in response to hormonal changes triggered by increasing sun exposure. Even while snow remains in some regions, life prepares to expand.

People living centuries ago learned to read these transitions without scientific instruments. They felt changes in the rhythm of the day, the scent of wind and the behavior of animals. The alignment of Setsubun with the solar calendar reflects this intuitive knowledge. The festival celebrates not the arrival of spring but the direction toward it.

Reflection and Ritual for Setsubun

Choose a quiet evening near Setsubun to mark your own threshold between seasons. Hold a handful of dried or roasted beans. Stand at your doorway and breathe deeply. Think about what you wish to release. Worry, heaviness, cluttered thoughts, or habits you are ready to leave behind.

Gently toss a few beans outward and speak a phrase of your choosing that carries the same spirit as Oni wa soto, fuku wa uchi. It may be as simple as trouble out, blessing in. Then take a step inside your home and place a few beans near the threshold. Imagine them as seeds of health and good fortune.

If you have access to holly or another evergreen branch, place a single leaf above the door for protection. Allow it to remind you that resilience grows even in cold seasons and that spring begins not with warm weather, but with the movement of light.

4 / LUNAR NEW YEAR: BLOSSOMS OF GOOD FORTUNE (CHINA AND DIASPORA COMMUNITIES WHEN IT FALLS IN FEBRUARY)

When Lunar New Year falls in February, the month becomes bright with color, tradition and anticipation. Families gather across China and throughout global Chinese communities to honor one of the most significant seasonal celebrations in the world. Homes are filled with red decorations, tables with symbolic foods and doorways with greetings meant to welcome prosperity, joy and renewal.

Lunar New Year does not mark the warmth of spring. Instead it celebrates a deeper shift in the year. It arrives when winter still lingers, yet its customs acknowledge that new energy is on the rise. Daily life slows, families reconnect and rituals of cleansing and blessing prepare homes for the year ahead. Above all, the festival honors a long tradition of blending myth with observation, folklore with agricultural rhythms and celebration with reflection.

Renewal Rituals and Monster Lore

One of the most enduring stories associated with Lunar New Year is the legend of Nian. According to this lore, Nian was a fearsome creature that emerged at the end of the year to wreak havoc on villages. People discovered that Nian feared loud sounds, bright colors and the crackling pop of fire. To protect themselves, they hung red decorations and set off firecrackers. Over time, these practices became joyful traditions rather than acts of fear.

The story of Nian reflects a universal theme found in many winter rituals. At the boundary between seasons, people confront lingering uncertainties or troubles, then mark the transition with symbols of strength and renewal. Red became the color of protection, courage and celebration. Drums, fireworks and lively gatherings banished the quiet heaviness of winter.

Other renewal customs included offering prayers, visiting ancestors, wearing new clothing and preparing foods that symbolized longevity, unity and prosperity. These actions helped anchor families in hope as they stepped into the new year.

Red is believed to attract good luck, protect against harm and encourage joy. During Lunar New Year, doorways, banners and envelopes shine in vivid shades of red to welcome blessing into every corner of the home.

Cleansing the Household

Household cleansing is one of the most meaningful practices associated with Lunar New Year. Before the first day of the celebration, families sweep their homes, wash linens, dust high shelves and tidy every room. Cleansing is done not only for physical cleanliness but also for symbolic renewal. It creates an open space for good fortune to enter.

During the celebration itself, however, sweeping is avoided. The belief holds that sweeping on the first days of the new year may remove newly arrived luck. Instead, people enjoy their refreshed surroundings and focus on welcoming abundance rather than removing it.

This practice mirrors the natural world. February contains a fullness of preparation. Just as plants ready themselves for budding, homes and hearts prepare for a cycle of prosperity. Cleansing clears the mind as much as the living space.

LEAVING LAST YEAR BEHIND

Old brooms were sometimes thrown away before the new year began so their worn bristles would not carry misfortune into the next cycle. Clean tools symbolized a clean start.

Plant Focus - Plum Blossoms and Citrus

Among the many symbols associated with Lunar New Year, plum blossoms and citrus fruits hold a special place.

Plum blossoms represent endurance, courage and the arrival of hope in cold seasons. They bloom before most other flowers and often open while frost still coats the ground. Their slender branches carry clusters of soft petals, each bloom a quiet promise of renewal. In art and poetry, plum blossoms stand for resilience and purity.

Citrus fruits such as tangerines, mandarins and oranges are displayed in homes and given as gifts. Their bright, warm colors evoke sunlight and abundance. In Chinese language and symbolism, the words for certain citrus fruits sound similar to terms for luck and wealth, amplifying their meaning during the new year.

Both plants reflect the heart of the celebration. Plum blossoms honor the endurance required to move through winter. Citrus honors the joy and prosperity hoped for in the year ahead.

Ecology - Why Some Trees Bloom in Late Winter

Early blooming trees like the plum have evolved to take advantage of the quiet window between winter and spring. During this time there is less competition for pollinators, and cooler temperatures help protect delicate blossoms from quick decay. These trees store energy throughout the dormant season and release it once daylight begins to grow.

Bud formation begins long before blooms appear. Even in the coldest weeks, sunlight triggers a rise in certain growth hormones within the branches. This early awakening allows plum blossoms to open before many other species. Their blossoms can tolerate chilly nights better than most flowers, which is why they thrive when snow still lies on the ground.

> **Plant and Ecology Notes:**
>
> **Plum Blossoms**
> Prunus mume
>
> • One of the earliest flowering trees in East Asia
> • Known for blooming in cold conditions
> • Symbol of perseverance and renewal
> • Featured heavily in poetry, painting and seasonal festivals
>
> **Citrus Fruits**
> • Evergreen trees that retain foliage through winter
> • Bright colors symbolize wealth and light
> • Often included in offerings and displayed in pairs for balance

Citrus trees follow a different rhythm. As evergreens, they maintain their leaves through winter and respond to increasing sunlight with slow internal shifts. Although they do not bloom in February in most regions, their fruits ripen during the cooler months, allowing their vibrant colors to shine during the new year season.

Nature teaches that renewal often begins beneath the surface. Long before warmth returns, trees and plants align themselves with the growing light.

Reflection and Ritual for Lunar New Year

On an evening close to the new year, pause and create a small clearing in your home. Sweep the space gently, then rest the broom aside. Place a branch of blossoms or a simple flower in a vase. Add a citrus fruit next to it. Let these natural symbols represent renewal, courage and the promise of joy.

Think about what you want to welcome into the coming year. Write a word or phrase that carries this intention and place it beneath the citrus or beside the vase. Light a candle nearby and sit for a moment with the soft glow.

If you feel called to do so, gently tap the table or floor with your fingertips. This small gesture mirrors the tradition of encouraging good energy to enter the home. Allow yourself to feel gratitude for the year that has passed and hope for the one beginning to unfold.

5 / GROUNDHOG DAY AND ANIMAL WEATHER WISDOM (NORTH AMERICA)

In North America, the second of February carries a tradition both playful and surprisingly ancient. Groundhog Day, known for predicting the length of winter based on the appearance of a small weather watching animal, is more than a modern curiosity. Beneath its lighthearted surface lie stories carried from Europe, shaped by agricultural cycles and old beliefs about animals as messengers of seasonal change.

Although most people associate February second with a single groundhog in Pennsylvania, communities across Europe once looked to bears, badgers or hedgehogs to read the future of the weather. When settlers crossed the Atlantic, they carried these customs with them. Lacking the same animals, they turned to creatures native to the new land. The groundhog, known for its deep winter dormancy and early emergence, became the chosen forecaster.

Despite the humorous note of modern celebrations, Groundhog Day preserves a deeper truth. People have long watched animals for signs that winter was shifting. In times before scientific instruments, observing wildlife was one of the most reliable ways to predict upcoming conditions. These patterns helped farmers decide when to plant, how to care for livestock and when to expect the earth to soften beneath their feet.

Ancient European Roots of Animal Omens

Long before groundhogs took on the role of seasonal clairvoyants, European communities believed that certain animals possessed insight into the coming weather. Badgers, hedgehogs and bears were particularly watched, since their habits shifted with seasonal cycles. If the animal emerged from its den on a clear day and retreated quickly, people believed more winter was coming.

If clouds softened the sky, spring was thought to be near.

These traditions developed from an understanding of wildlife behavior. People noticed that animals responded to changes in light, temperature and food availability. Over generations, these observations became woven into folklore. When immigrants settled in North America, they adapted the custom to the animals around them. The groundhog, with its strong hibernation cycle and early February stirring, fit the role perfectly.

While the European stories have faded in many regions, their echoes remain within the North American custom. The belief that an animal's behavior can reveal hidden knowledge reflects a respect for nature that shaped daily life for centuries.

FOLKLORE OF FEBRUARY / 29

Shadow Signs

If the day is bright and the animal sees its shadow, winter continues. If the sky is cloudy and no shadow appears, spring will arrive soon. This simple image became one of the most widely recognized pieces of seasonal folklore in North America.

February Shadows and Seasonal Forecasting

Shadows play a central role in Groundhog Day. A clear day casts long, sharp shadows in early February, signaling the supposed return of winter's strength. Cloudy skies soften shadows or hide them completely, suggesting the early arrival of spring.

While the lore is framed playfully today, it reflects a genuine interest in seasonal timing. Communities needed to know when to begin planning for planting, when livestock could graze safely and when travel would improve. Observing the weather on a specific day allowed families to make predictions based on patterns passed down through generations.

The shadow tradition also mirrors Candlemas weather lore found in Europe. The two customs share more than a date. Both interpret early February light as a sign of the winter ahead.

> **Weather Lore Insight**
>
> Bright sun in early February often means colder nights, since clear skies allow heat to escape the surface of the earth. Cloudy days can signal milder conditions. This is why weather lore sometimes aligns loosely with real patterns, even though it is not a scientific predictor.

Plant Focus - Burrow Plants and Early Shoots

Though Groundhog Day is centered on animals, the landscape around burrows tells its own story.

Plants that grow near burrow entrances often reveal the earliest signs of seasonal change.

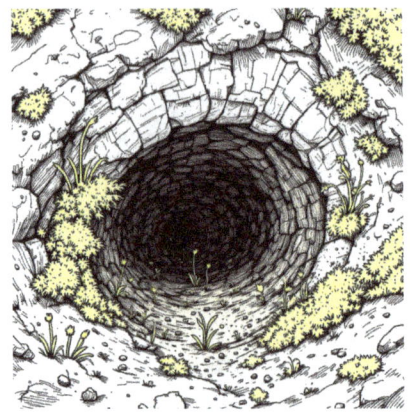

Groundhogs clear soil regularly, allowing light to reach the surface. This disturbance invites resilient plants to sprout early.

Certain species like chickweed, coltsfoot and hardy winter annuals can be found emerging near den sites. Mosses brighten with moisture, gaining a deeper green as daylight increases.

> *Even in cold climates, burrow areas warm slightly due to soil turnover, making them natural microhabitats for early plant growth.*

These first shoots remind us that nature prepares for renewal long before spring is visible. While the groundhog captures public attention, the plants at its feet whisper the same message.

Science - Animal Behavior and True Weather Patterns

Groundhogs, like many hibernating animals, rely primarily on internal biological rhythms rather than daily weather to determine when to wake. Their emergence is triggered by hormonal changes linked to increasing light and shifts in body temperature. They do not consciously predict the weather, yet their timing does reflect seasonal patterns that humans once read carefully.

> **Plant Notes**
> Early Ground Level Growth
>
> **Chickweed**
> (Stellaria media)
> • One of the earliest green plants to appear
> • Thrives in disturbed soil near burrows
> • Often stays green beneath snow
>
> **Coltsfoot**
> (Tussilago farfara)
> • Produces bright yellow flowers before leaves appear
> • Prefers sunny, well drained soil
> • Symbol of early hope in many traditions
>
> **Mosses**
> • Respond quickly to moisture and light
> • Serve as indicators of subtle seasonal shifts

Scientists note that groundhog predictions are not accurate for long term forecasting. However, the broader concept of reading wildlife for seasonal clues remains meaningful. Many animals respond more quickly to environmental changes than humans do. Birds alter their calls as daylight grows. Foxes begin mating. Trees show slight bud swelling. Soil communities stir beneath the surface.

The instinctive behaviors of animals reveal the same truth as traditional festivals. February is a month of preparation. Winter is still present, but the earth has begun its slow turn toward spring. Observing wildlife at this time offers a glimpse into the natural processes that shape the year.

Reflection and Ritual for Groundhog Day

Find a quiet outdoor place or window view on the second of February. Notice the quality of light. Observe whether shadows fall long and sharp or soft and muted. Watch for small movements in the natural world, such as birds calling or early green patches emerging.

Hold these observations gently and consider what they mean for you. Are you still in a season of rest, or do you feel the first signs of awakening inside yourself? Write down a single insight or intention connected to this moment.

If you wish, place a small stone or seed near your doorway to honor the earth's quiet renewal. Let it remind you that change often begins with the smallest signs, whether in an animal's emergence or a single green shoot pushing through cold soil.

6 / CARNIVAL: WINTER'S MASK AND MISCHIEF (GLOBAL TRADITIONS)

ACROSS MANY CULTURES, the weeks before spring arrive with color, laughter and a hint of wildness. Carnival, celebrated under various names throughout Europe, Latin America and parts of Africa, brings a burst of energy into the final stretch of winter. Though each region expresses the festival differently, the core themes remain constant. People feast, disguise themselves, turn the ordinary world upside down and prepare for the transition into a new season.

> *Carnival is a celebration of human spirit in the coldest time of year.*

It is a release valve for winter's restraint, a moment when joy becomes nourishment, costumes become stories and communities come together to remind themselves that warmth and abundance are approaching. Beneath the music and dancing lies a deep seasonal wisdom. Carnival honors the last days of scarcity while simultaneously welcoming the lengthening light and the promise of renewal.

Role Reversals and Late Winter Feasting

Role reversal has been a defining feature of Carnival since its earliest forms. In medieval Europe, servants dressed as nobles, nobles disguised themselves as commoners and communities temporarily inverted their social hierarchies. Laughter, mischief and playful disruption allowed people to release tension built up through long winters. These symbolic flips gave communities a safe space to express humor, criticism and relief before returning to daily routines.

Feasting also plays a central role. Carnival arrives before fasting periods in many traditions, making it a time to use remaining grains, fats and stored ingredients. In cold climates, food stores grow thin by late winter.

> *Carnival meals strengthen both body and spirit with rich, warming dishes.*

Pancakes, doughnuts, buns, dumplings and braided breads appear on tables as reminders of the sweetness and abundance that will return in spring.

In warmer regions, the festivities take on even brighter forms with processions, music, dancing and elaborate costumes. Whether in Venice, Rio de Janeiro or New Orleans, Carnival speaks the same message. Celebrate now, for the world is about to shift.

FOLKLORE OF FEBRUARY / 35

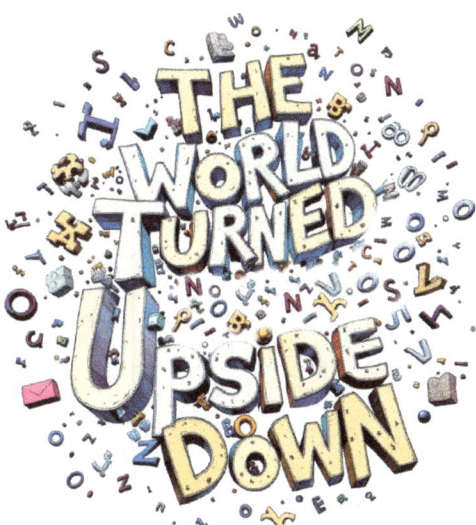

Carnival allowed temporary suspension of rules. Masks and costumes made all people equals, symbolizing a brief return to the freedom and wildness of earlier ages.

Symbolism of Masks

Masks are the heart of Carnival. They conceal identity while revealing deeper truths. When a person wears a mask, they step outside the expected roles of daily life. They become part of a shared story, a moment of collective play and imagination.

In Venice, masks once allowed people from different social classes to mingle freely. Elaborate designs with soft colors, feathers and gilded shapes transformed wearers into characters from myth or dream. In Brazil, masks and costumes carry African, Indigenous and European influences, each expressing joy, resilience and cultural memory. Masks symbolize transformation in many spiritual and communal traditions.

They remind people that identity is fluid and that renewal requires stepping into a different perspective. During Carnival, masks invite people to leave behind the heaviness of winter and embrace creative freedom.

Mask Meanings Across Cultures

Venetian Masks
• Conceal identity, allowing freedom of expression
• Often feature gold details and elongated shapes
• Symbolize elegance, mischief and mystery

Latin American Masks
• Bright colors and bold patterns
• Represent spirits, ancestors or cultural archetypes
• Celebrate diversity and resilience

African Influenced Masks in Carnival
• Honor heritage and communal storytelling
• Feature geometric designs and symbolic colors

Material Focus Dyes, Grains and Spices

Carnival traditions rely on materials that carry both practical and symbolic meaning.

Dyes - Bright colors appear in masks, costumes and decorations. Natural dyes historically came from plants, minerals or insects. Indigo, saffron, madder root and cochineal brought life to fabrics during a season dominated by gray skies.

Grains - Wheat, rye and barley stored from autumn transform into breads and pastries during Carnival.

These grains symbolize sustenance during the final stretch of winter scarcity. Their use in celebratory foods reflects the human desire to honor what remains before the next harvest cycle begins.

Spices - Cinnamon, nutmeg, ginger and cloves brighten winter dishes. Spices warm the body and enliven the senses. In many regions, Carnival pastries and breads rely on spices to create a comforting aroma that lifts the spirit.

> **Material Notes:
> Late Winter Foods**
>
> **Wheat and Barley**
> • Foundation of breads and pastries
> • Provide warmth and energy
> • Symbolize endurance during winter
>
> **Cinnamon and Nutmeg**
> • Stimulate circulation
> • Associated with prosperity and comfort
> • Common in sweet carnival dishes
>
> **Natural Dyes**
> • Indigo for blue
> • Cochineal for red
> • Saffron for gold
> • Bring vibrant tones to costumes and banners

Together these materials show how communities use creativity and resourcefulness to bring color and flavor into the darkest days of the year.

Biology - Why Bodies Crave Richness in Late Winter

Carnival feasting reflects more than tradition. Human biology responds to seasonal change in ways that influence appetite and energy. During winter, colder temperatures and shorter daylight hours affect metabolism, mood and hormonal balance.

Lower light levels increase melatonin production, which can lead to low energy or sluggishness. Bodies crave calorie dense foods to maintain warmth and fuel daily functioning. Fats, grains and spices provide steady energy and comfort. These cravings are not simply indulgence. They are rooted in ancient survival needs.

By late winter, people historically relied on stored foods. Feasts gave them strength for the final stretch before spring crops emerged. Even today, the desire for warm, rich dishes during this time persists. Carnival celebrates these instincts rather than suppressing them. It allows communities to nourish themselves while honoring seasonal cycles.

Reflection and Ritual for Carnival

 Choose an evening to honor the spirited nature of Carnival. Place a simple mask or fabric scrap on a table. Light a candle nearby. Think about the roles or habits you have carried through winter and consider which ones may be ready to loosen.

 Prepare or enjoy a small feast item such as bread, fruit or a warm spiced drink. Let the aroma and flavor remind you of the body's need for comfort during cold seasons. Take a moment to imagine yourself stepping into a new role or perspective as the light grows.

 If you feel playful, create a quick drawing of a mask or craft one from paper. Allow it to express a part of yourself that wishes to be seen or transformed. Keep it as a reminder that joy, creativity and renewal often begin with a single imaginative spark.

7 / LOVE, LOTS AND DIVINATION: MIDWINTER COURTSHIP BELIEFS (EUROPE AND ELSEWHERE)

Long before modern valentine cards, heart shaped confections or romantic marketing, February held a quiet but powerful association with love and courtship. In many parts of Europe, the midpoint of winter was believed to influence relationships, destiny and emotional renewal. These beliefs grew from a blend of observation, hope and the rhythms of the natural world. They remind us that love, like the seasons, has always been shaped by cycles of anticipation, awakening and subtle change.

The earliest February love traditions were not about grand gestures. They were about signs, omens, pairings and the tender awareness that life was beginning to stir beneath cold ground. People watched the sky for returning birds. They noted which flowers survived frost and emerged early. They relied on playful forms of divination to predict the shape of relationships. Through these customs, communities expressed both longing and optimism during a time when winter seemed ready to loosen its hold.

. . .

Valentine Folklore Before the Holiday Became Commercial

The origins of Valentine traditions are layered with history and myth. In medieval Europe, February was considered the month when nature prepared to awaken. People believed this made it an ideal time to ask questions about love. Long before printed cards, young people exchanged small handmade tokens, poems or charms to express admiration.  These early valentines were simple and heartfelt.

One enduring belief claimed that the first person of the opposite sex one saw on Valentine's morning would be the partner they would marry. Another tradition involved drawing lots with names written on them. Young people gathered, pulled names from a container and became symbolic partners for the upcoming season. These pairings were often playful, although some were believed to predict real outcomes.

Many customs blended romance with protective magic. A woman might place bay leaves under her pillow to dream of her future partner. A man might carry a sprig of certain flowers to attract love or good fortune. These gestures were rooted in a belief that nature's stirring energy could guide human hearts as well.

St. Valentine's Birds

One medieval English belief held that birds chose their mates on February fourteenth. This idea spread quickly and influenced the human association between Valentine's Day and courtship.

Bird Marriage Myths

Birds have long symbolized love, loyalty and partnership. Many cultures believed that birds began pairing in midwinter as the light lengthened. People observed birds calling to one another more frequently and interpreted these changes as acts of courtship.

Different species became tied to specific romantic meanings. Doves symbolized devotion and gentleness. Robins represented passion and warmth. Sparrows embodied humble, lasting love. Seeing certain birds on Valentine's morning was considered a sign of what kind of partner one might find. A goldfinch meant wealth and stability. A robin meant a faithful but lively union. A sparrow suggested a simple and contented life.

Bird behavior reinforced these ideas. Even in winter, many species begin to vocalize in ways that strengthen pair bonds. Their calls and movements provided the inspiration for countless tales of loyalty and affection.

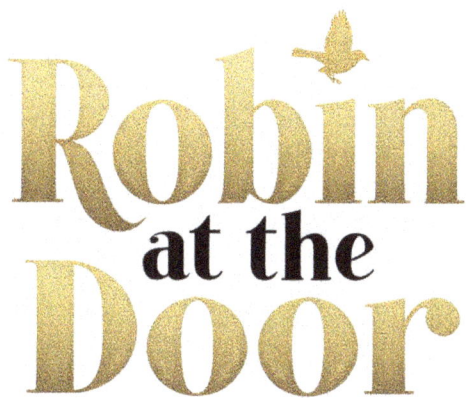

Seeing a robin tapping at a window in February was once believed to foretell a love story soon to unfold.

Plant Focus - Violets and Rose Lore

Winter offers few flowers, so those that appear in or near February gained powerful romantic meaning.

Violets, often among the earliest flowers to bloom in mild regions, symbolize modesty, truth and quiet affection. Their deep purple petals were linked to sincerity and spiritual love. In folk customs, wearing violets could draw honest affection or bring clarity to matters

of the heart. Their fragrance was used in early perfumes meant to inspire tenderness.

Roses, especially red and pink varieties, carry global associations with love. While roses do not bloom naturally in February in most climates, dried petals, essences and preserved wreaths were used in midwinter rituals. Red roses became symbols of passionate love, while white roses represented purity and devotion. In some traditions, rose water was sprinkled in rooms to bless partnerships or soothe emotional tensions.

Both flowers speak to the longing for beauty during cold months. They represent love that endures despite the harshness of winter.

Science - Early Bird Mating Songs and Reproductive Triggers

Birds are among the earliest creatures to respond to February's changing light. While people once interpreted their calls as romantic gestures, science reveals a deeper process at work. As daylight increases, light receptors in a bird's brain trigger hormonal changes. These hormones influence behavior, encouraging vocalization, territory marking and bonding.

> **Plant Lore Notes:**
>
> **Violets**
> (Viola odorata)
> • Associated with modesty and truthful love
> • Among the earliest sweet scented flowers
> • Used in charms for clarity of heart and intention
>
> **Roses**
> (Various species of Rosa)
> • Symbolize passion, purity or friendship depending on color
> • Frequently featured in midwinter love rituals
> • Essence used to soothe, bless or inspire affection

Birds begin to sing more often not due to temperature but because their biological clocks sense the growing day. These early songs help strengthen existing pair bonds or attract new partners. In many species, males begin practicing calls weeks before mating begins. Their songs declare readiness and vitality.

This behavioral shift mirrors the folklore surrounding midwinter love.

Communities observed the natural world awakening and drew connections to human emotions. The pairing of birds became a symbol of what people hoped to find in their own lives. Science supports the idea that February is a month of preparation, signaling that life is preparing to blossom.

Reflection and Ritual for Midwinter Love

Set aside a quiet moment during February to explore your own understanding of love, companionship or connection. Hold a violet or another small flower, or use an image of one if none are available. Sit by a window at dawn or dusk when the light feels gentle.

Think about what qualities you value in love or friendship. Write down a word that represents the type of connection you wish to welcome or strengthen. Fold the paper and place it beneath a small stone, flower or candle.

Listen for any bird calls outside. These early songs remind you that nature prepares for renewal even in winter. Let that sense of awakening guide your thoughts toward the relationships you want to nurture in the coming season.

PART 2
PLANTS, TREES AND MATERIALS OF FEBRUARY

8 / SNOWDROPS
LANTERNS AT GROUND LEVEL

SNOWDROPS APPEAR when the world still seems undecided. Winter has not released its grip, yet the land no longer feels entirely asleep. Frost still rims the edges of fields and paths. The soil is cold to the touch. And yet, from beneath this restraint, snowdrops rise with calm determination.

> *They do not announce themselves loudly. They bow their heads, lift pale bells toward the light and bloom while snow may still lie around them.*

For centuries, people have noticed this quiet courage. Snowdrops became symbols not of triumph but of endurance. They are not flowers of celebration or abundance. They are flowers of persistence. In folklore, they mark the exact moment when winter loosens just enough to let hope through.

. . .

Mythic Purity and Protection

Snowdrops carry a long history of symbolic meaning across Europe. Their white petals have been linked to purity, humility and protection. Unlike later spring flowers that open wide and face the sun, snowdrops keep their heads bowed, as if in contemplation. This posture inspired associations with modesty and reverence.

In some traditions, snowdrops were believed to guard households during the turning of the season. A bloom placed near a threshold was thought to protect against illness and misfortune. This belief tied the flower to liminal spaces. Doorways, windowsills and garden edges became places where snowdrops quietly stood watch.

There are also gentler legends attached to the flower. One story tells of snowdrops appearing as a gift of comfort after a long winter, offering reassurance that warmth would return. Another suggests they were formed from drops of light that fell to earth when winter began to lift. These myths reflect an emotional truth. When food stores were low and days were still short, a single white flower could shift the spirit of a household.

Snowdrops were not always freely gathered. In many regions, people believed that bringing too many indoors could invite sorrow. This caution speaks to the respect people held for the plant. Snowdrops were meant to be noticed, not claimed. Their power lay in their presence, not in possession.

Biology of Frost Defiance

The ability of snowdrops to bloom in winter conditions is not accidental. It is the result of precise biological adaptation shaped over thousands of years. Snowdrops are geophytes, meaning they grow from underground bulbs that store energy during warmer months. This stored energy allows them to push upward before other plants awaken.

One of the most remarkable traits of snowdrops is their ability to generate heat. Through a process known as thermogenesis, snowdrops can raise the temperature of the tissues surrounding their stems. This allows them to melt snow directly above the bulb and carve a small passage to the surface. It is a quiet act of persistence, powered by careful preparation rather than force.

Their structure also protects them from damage. The nodding shape of the flower shields pollen from frost and rain. The waxy coating on the leaves helps prevent moisture loss and cold injury. Snowdrops grow close to the ground where temperatures are often more stable than the air above.

Light plays a critical role in their timing. Snowdrops respond strongly to increasing day length rather than warmth. Even when temperatures remain low, the lengthening days of February signal that it is time to grow. This sensitivity to light connects them directly to the themes of the season. Snowdrops bloom not because winter has ended, but because the direction of the year has changed.

Lanterns at Ground Level

Snowdrops are often described as lanterns, and the image is an apt one. Their pale petals catch and reflect even the faintest light. In early morning or late afternoon, they seem to glow against dark soil and leaf litter. This visual contrast gives them a presence far greater than their size.

The lantern image carries symbolic weight.
Lanterns guide travelers, mark thresholds and offer reassurance in darkness. Snowdrops do the same for the natural world. They appear at a moment when the landscape feels

suspended between seasons. Their glow suggests that something beneath the surface is ready to rise.

In this way, snowdrops become teachers of patience. They do not rush the season forward. They simply show what is already happening quietly underground. Roots are stirring. Buds are swelling. Birds are adjusting their songs. Snowdrops stand at the meeting point of stillness and movement.

Observing the First Flowers

To encounter a snowdrop is to be invited into a slower way of seeing. These flowers are easy to overlook if one moves too quickly. They grow low and often appear in shaded places. Old paths, woodland edges, churchyards and quiet gardens are among their favored homes.

When you find one, pause. Notice how the flower bends rather than stands upright. Observe the texture of its petals and the green markings hidden inside the white. Look at the ground around it. Often there are signs of life nearby. Moss brightening with moisture. Tiny shoots just beginning to push upward. Soil loosening beneath frost.

Snowdrops rarely grow alone. They tend to appear in small groups, as if reminding us that endurance is strengthened by community. Their clustered presence transforms a bleak patch of ground into something gently alive.

Observation rather than collection honors their role in the season. In many places, snowdrops are protected and should not be picked. Their power lies in their timing and setting. They belong to the moment when winter yields just enough to let light through.

Science and Seasonal Meaning

From a scientific perspective, snowdrops are indicators of phenological change. Phenology is the study of how living organisms respond to seasonal cues such as light and temperature. Snowdrops are among the earliest biological signals that winter is shifting.

Their bloom does not mean spring has arrived. Instead, it confirms that the cycle has turned. This distinction is important. February is not about arrival. It is about direction. Snowdrops embody this truth. They bloom because the light has changed, not because the cold has disappeared.

This aligns closely with the folklore surrounding them. People sensed that snowdrops marked a boundary rather than a destination. They were never symbols of completion. They were symbols of passage.

Reflection and Practice - Lanterns in the Cold

During February, take a slow walk with the intention of looking downward rather than outward. Seek places where the ground meets paths, walls or trees. Look for small signs of light close to the earth. A snowdrop, a patch of frost reflecting the sky, a pale stone catching the sun.

When you find such a moment, stand quietly with it. Let it remind you that growth does not always announce itself. Sometimes it arrives bowed, quiet and persistent.

If you wish, write down one small change you have noticed in your life or surroundings this season. Keep it simple. Like the snowdrop, it does not need to be large to matter. It only needs to signal that the light has begun to return.

9 / HAZEL
WISDOM AT THE WOODLAND EDGE

Hazel has long been regarded as a tree of insight, intuition and hidden knowledge. Growing at the edges of forests, along riverbanks and near old paths, hazel often stands where boundaries meet. In folklore, places of transition hold power, and the hazel tree reflects this truth with its presence in stories of wisdom, divination and protection.

February is a subtle month in the woodland world. The forest does not yet bloom, yet its quiet movements signal what is to come. Hazel is among the first to reveal the shift. Its catkins lengthen in the pale sunlight, releasing pollen into the cold air even while the ground remains hard beneath them. These slight changes were noticed by earlier communities who understood that hazel was a tree that spoke in quiet signs.

Hazel invites us to slow down, observe and listen. It teaches that knowledge does not always arrive loudly. Sometimes it appears in a single tassel of gold, trembling in a breeze that still carries a winter chill.

Rods, Nuts and Divining Traditions

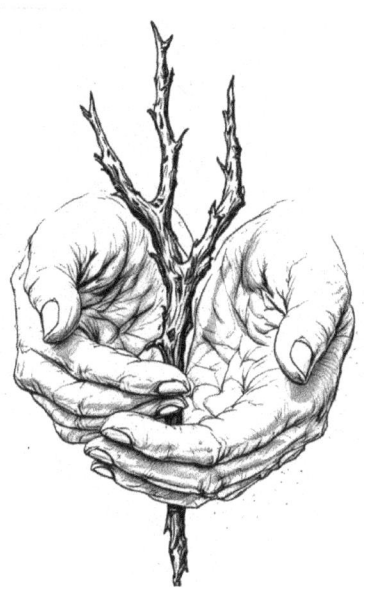

Hazel rods have been used for centuries in divining practices throughout Europe.

Whether a person was searching for water, minerals or subtle shifts in the earth, hazel was considered a receptive and trustworthy wood.

The forked rod, often cut from a young branch, was believed to respond to unseen influences and guide the seeker.

This practice, known as dowsing or water witching, required both sensitivity and patience. Hazel was chosen because it symbolized insight and because its flexible wood responded well to slight movements. Even for those who did not practice divination, the hazel rod represented the search for understanding and the desire to connect with the land.

Hazel nuts also carried strong associations with wisdom. In Irish and Scottish lore, salmon gained knowledge by eating hazel nuts that fell into sacred pools.

These stories emphasized the idea that wisdom comes from nature and that hazel stands

> **Hazel and the Salmon of Wisdom**
>
> In Irish legend, nine hazel trees grew around a sacred pool. When their nuts fell into the water, salmon ate them and gained deep knowledge. Whoever ate such a salmon would inherit this wisdom. The tale highlights hazel as a tree of insight and spiritual nourishment.

near the source of that knowledge. Families often kept hazel wood in their homes for protection. A small piece placed above a doorway or near a cradle was believed to guard against harmful unseen forces. The tree offered a sense of comfort and watchfulness.

Catkins and Wind Pollination

Hazel catkins are among the first visible signs of life returning to the forest. In February, the male catkins lengthen into tassels that sway with the wind. Their warm golden color contrasts beautifully with the otherwise muted tones of late winter.

Unlike many flowering trees, hazel relies on wind rather than insects for pollination. The movement of air carries pollen from the dangling male catkins to the tiny, bright red female flowers that sit almost hidden along the branches. These blossoms are easy to miss unless one looks closely. Their ruby colored threads appear like delicate sparks of warmth in a cold season.

This pollination method mirrors the subtle nature of February. Transformation happens quietly, carried on invisible currents. Hazel teaches that important change often begins with small gestures, soft movements and gentle exchanges rather than dramatic displays.

Ecological Role

Hazel thickets offer shelter to birds and small mammals. Their early catkins feed pollinating insects once temperatures rise, and their nuts support species such as squirrels, dormice and jays.

Quiet February Forests

February forests have a particular quietness. Footsteps sound crisp on frosted leaves. Branches stand in patient silhouettes. Yet this apparent stillness holds a delicate energy. Look closely and you will see signs of waking life.

Hazel stands out in this landscape. Its catkins shimmer in the low light, catching sun that slips between branches. Birds pause on its limbs and test early calls.

Botanical Notes on Hazel
(Corylus avellana)

• Grows in woodlands, hedgerows and forest edges
• Male catkins release pollen in late winter
• Female flowers appear as tiny red tufts
• Nuts form later in the year and provide valuable food for wildlife

Small prints in the snow reveal animals seeking food near its base. Even the air smells slightly different near hazel stands, carrying a hint of sap that prepares to rise.

Walking through a February forest invites a slower pace. The mind softens in the hush, and the senses open. Hazel encourages this awareness. It draws attention to detail. To the way sunlight falls. To the structure of a branch. To the pause between breaths. This is the wisdom of the woodland edge. Presence, observation and patient trust.

Reflection and Ritual for Hazel

Find a quiet outdoor place or a comfortable indoor space with a view of trees or earth. Hold a small stick or twig, preferably from a fallen branch, as a symbolic hazel rod. Close your eyes and breathe gently.

Ask yourself what guidance you seek. Not in words or demands, but in openness. Let your awareness rest in your hands. Notice any feeling of connection or calm. When ready, place the twig near a window or doorway where it can receive light.

If you are able to walk through a woodland or park in February, search for hazel catkins. Spend a moment observing their form. Quietly acknowledge the wisdom of small beginnings.

10 / WILLOW

FLEXIBILITY, FLOW AND EARLY GROWTH

WILLOW IS a tree that carries the quiet rhythm of water and the early pulse of spring. Found along riverbanks, ponds and wetlands, willow thrives where land meets flow. Its branches bend without breaking, its roots drink deeply and its leaves move in patterns shaped by wind and current. These qualities made willow a symbol of resilience, flexibility and healing across many cultures.

In February, while most trees appear rigid in winter stillness, willow begins to shift. Its branches hold a faint flush of green. Its buds swell in response to the increasing light. Even the texture of the bark feels different.

Willow senses spring long before other plants dare to. It stands at the threshold of seasons, whispering soft promises of growth.

Folklore of Waterways and Spirits

Waterways have always been places of mystery where the physical and unseen worlds seem to mingle. Willow, growing along these edges, became associated with spirits, guardians and healing forces. I

n European folklore, willows were said to shelter gentle water beings who protected travelers or guided those who sought emotional healing. In some traditions, willows bowed over rivers as if listening to stories carried downstream.

Willow branches often appeared in rituals meant to soothe grief or calm troubled hearts. Their graceful movement in a breeze reminded observers that emotions, like water, flow rather than stay fixed. The tree's ability to bend without snapping influenced its symbolic meaning as well.

Willow taught people that strength could be found in yielding, that survival depended not only on endurance but on adaptability.

Certain stories warned against falling asleep under a willow at dusk because the spirits of the place might whisper dreams too deep for comfort. Other tales described willows as wise watchers of the land, trees that held memories of those who had passed. Whether benevolent or uncanny, willow was always seen as a tree in touch with forces just beyond human sight.

Its connection to water made it a companion to those who practiced healing arts. Willow bark was known for its pain relieving properties, and some healers believed the tree's knowledge flowed through its sap. To sit beneath a willow was to seek calm, clarity and emotional renewal.

> **Willow for Healing**
>
> Across Europe, willow branches were placed near bedsides to ease grief and calm restless thoughts. The tree's bending form inspired the belief that emotions could soften and move again.

. . .

How Willows Respond to Subtle Shifts in Light

Even before winter loosens its hold, willows begin to change. They respond quickly to the growing length of daylight. The increase in sun activates hormones within the tree, causing buds to swell and branches to take on a brighter hue. While other trees remain dormant, willows prepare for growth with remarkable sensitivity.

The pencil thin twigs of a willow reveal this early awakening most clearly. In February they carry hints of gold, green or red depending on the species. This shift is subtle bu

t noticeable to anyone who pauses and looks closely. The bark becomes more pliant, the branches more responsive to touch. Inside the buds, tiny leaves are forming long before spring warmth arrives.

Willows are well suited to this early preparation because they thrive in moist soil. Water in the ground helps support the metabolic changes necessary for new growth. Even during cold nights, willow roots work steadily. They absorb moisture and minerals that fuel the tree's early buds.

Botanical Notes on Willow
(Salix species)

• Thrives in wetlands, riverbanks and damp meadows
• Responds early to increasing daylight in late winter
• Produces flexible branches ideal for weaving
• Bark contains compounds once used for pain relief

Ecological Role
Willows provide early nectar for insects when spring begins. Their roots stabilize riverbanks, and their branches shelter birds.

. . .

Their responsiveness to light symbolizes clarity and intuition in many traditions. Just as the willow senses what is coming, people looked to the tree as a guide for trusting inner knowing. The tree's early shift became a natural metaphor for awareness and readiness.

Simple Weaving Rituals

Willow branches have long been used for weaving. Their flexibility makes them ideal for baskets, wreaths and symbolic objects created to mark seasonal transitions. In February, when willow twigs are supple but not yet busy with leaves, they are especially easy to shape.

Weaving with willow was traditionally a way to honor the coming of spring. People crafted small circles or knots to represent unity, protection or the cycle of renewal. Children learned to twist thin willow strands into simple hoops. Adults formed more intricate shapes for household use or seasonal celebrations

A common practice involved weaving a small ring from fresh willow and hanging it near a doorway. This ring symbolized flow, resilience and the ability to move gracefully through challenges. As the ring dried, it retained its shape, reminding the household that flexibility brings strength.

Another ritual involved creating a willow wand. A straight twig was stripped of small side branches and held during moments of reflection or meditation. Some believed that the wand helped direct healing energy, while others simply used it as a reminder to stay present, open and adaptable.

These weaving practices were not meant to be perfect or ornate. Their value came from the act of working with the tree, from feeling its smooth bark and listening to the quiet creak of its bending twigs. Willow weaving connected people to the natural world and to the promise of renewal that February brings.

. . .

Reflection and Ritual for Willow

Find a slender willow twig, or use a fallen one if you prefer not to cut from a living tree. Hold it gently and notice its flexibility. Let it remind you that strength can come from softness, and clarity from calm observation.

To create a simple weaving ritual, form the twig into a small circle. Overlap the ends and twist them lightly so the shape holds. This circle represents flow and resilience. Place it near a window or on an altar where it can catch the light.

Sit quietly for a moment and ask yourself where you might invite more flexibility into your life. What feels tight or rigid that could soften? What new growth is waiting for your attention?

As the days lengthen, allow the willow circle to remind you that early change often begins in subtle ways. Trust small awakenings. They lead to greater ones.

11 / EVERGREEN GUARDIANS
PINE, YEW AND JUNIPER

EVERGREEN TREES STAND through winter with a presence that feels both protective and ancient. While deciduous trees release their leaves and settle into deep rest, evergreens hold their color, their scent and their quiet strength. In February, when cold still dominates the landscape but light begins to stretch its way back into the world, evergreens offer comfort and continuity. Their branches stay full, their aromas warm the air and their folklore weaves through the histories of many cultures.

Pine, yew and juniper form a triad of winter guardians. Each carries distinct properties, both practical and symbolic. Pine speaks of cleansing and renewal. Yew whispers of endurance and ancestral memory. Juniper protects, purifies and clears the way for fresh beginnings. Together they create a protective landscape in late winter, bridging the fading darkness with the promise of spring.

Winter Amulets and Protective Properties

Across Europe and Asia, evergreens were viewed as vessels of life that refused to bow to the cold. People believed that trees capable of staying green during winter must hold special power.

Branches were hung in homes, placed over doorways or burned in small fires to cleanse the air. Long before air fresheners or modern disinfectants, the natural chemistry of evergreens provided both fragrance and genuine protection.

Pine branches were used to sweep negativity from households at the start of a new season. The needles crackled pleasantly when burned, and the scent of pine smoke was thought to ward off illness. Bundles of pine needles were sometimes hung near windows to invite clarity and bright energy during the dim months.

Yew carried heavier symbolism. Often found in churchyards and ancient burial sites, yew was connected to cycles of life, death and rebirth. While this might seem somber, yew's evergreen presence reminded communities that continuity endured through winter and beyond. Sprigs of yew were tucked into amulets for protection, remembrance or spiritual grounding.

Juniper was especially valued for purification. In many regions, people burned juniper sprigs to cleanse sickrooms or to protect newborns and travelers. Its sharp, invigorating scent symbol-

---

Juniper Smoke for Renewal

In many northern European regions, juniper smoke was wafted through rooms to chase away winter heaviness, illness and misfortune. People believed that juniper carried protective warmth into the coldest corners of a home.

ized renewed vitality. Juniper smoke was believed to repel harmful influences, both physical and unseen. During February, when illness was more common and households remained closed against the cold, juniper offered a sense of fresh air and protection.

Every evergreen used as a winter charm served a deeper purpose. These trees reassured people that life persists, even when the world lies in stillness.

> **Juniper for Safe Passage**
>
> Travelers once carried juniper berries in their pockets when crossing unfamiliar lands. The berries were believed to protect against misfortune and guide the traveler back home.

Chemistry of Aromatic Resins

The rich scents of pine, yew and juniper are not just pleasant aromas. They are complex chemical signatures developed by the trees to protect themselves. These aromatic resins contain compounds that discourage insects, inhibit the growth of microbes and help heal wounds in the bark. What humans perceive as cleansing or soothing fragrances are actually survival tools of remarkable sophistication.

Pine resins contain terpenes such as pinene, which create the sharp, refreshing scent associated with forests. These compounds have antimicrobial properties and help seal wounds, preventing decay. When pine needles or resin are warmed, these aromatic molecules fill the air with a scent often used to clear the mind.

Juniper contains juniperol and aromatic oils found in both the berries and needles. These oils are warming and stimulating. Historically they were used in tonics or burned to improve air quality in homes.

Juniper berries also provide flavor in culinary traditions and beverages. **Yew**, while less aromatic than pine or juniper, contains complex compounds that protect the tree from insects and grazing animals. Many parts of the yew are toxic, a natural defense that allows the tree to grow slowly and live for centuries. Though not used medicinally in folk practices as often as juniper or pine, yew was respected for its endurance and mystery.

These chemical properties subtly shape the role evergreens play in human culture. Their scents awaken the senses during winter months, lift moods, support clarity and evoke memories of forests where life continues even in the coldest days.

Evergreen Notes and Uses

Pine
• Needles used in teas and cleansing bundles
• Resin has antimicrobial qualities
• Symbol of clarity and bright energy

Yew
• Associated with ancestors and endurance
• Wood valued for ritual objects
• Evergreen color symbolizes continuity

Juniper
• Smoke used for purification
• Berries used for flavoring and protection charms
• Symbol of vitality and renewed strength

Late Winter Uses for Conifers

In late winter, conifers become especially valuable. Their needles, wood and berries provide materials for rituals, practical tasks and seasonal preparation.

Pine: In February, pine needles are gathered to make simple teas that carry a bright, restorative flavor. The vitamin rich infusion was once used to support health in months when fresh greens were scarce. Pine branches, woven into wreaths or bundles, symbolized cleansing and new beginnings as the household prepared for spring.

Juniper: Juniper branches were burned to refresh the stale air of winter homes. Berries were added to stews or preserved with meats, contributing both flavor and believed protective energy. In some traditions, juniper was used to bless tools or livestock as the agricultural year approached.

. . .

Yew: Though not used extensively for food or burning, yew wood was prized for crafting. Its density and smoothness made it ideal for small tools, ritual objects or bows. In February, a slender piece of yew might be carved into a token of remembrance or placed upon an altar to honor ancestors. Late winter is a season of preparation. Conifers supply what the land can offer at this time: fragrance, shelter, resilience and reminders of continuity. They help bridge the final weeks of winter with symbols of strength and endurance. Evergreens teach a simple truth. Even in the coldest months, not all life retreats. Pine, yew and juniper stand as guardians through the darkest part of the year.

Reflection and Ritual for Evergreens

Collect a small sprig of pine, yew or juniper from the ground, or use cuttings sold for crafting. Hold the sprig in your hands and breathe in its scent. Notice how the fragrance brings both clarity and calm. Place the sprig near a candle or a window. Reflect on what qualities evergreens represent for you. Endurance, protection, memory, or renewal. Take a moment to acknowledge which of these qualities you need most as winter draws to an end. If you feel called, create a small evergreen charm. Tie a few needles or berries together with thread and hang it above your doorway or near a desk. Let it serve as a guardian for the final stretch of winter and a reminder that strength can be quiet, steady and evergreen.

12 / MILK
AND THE QUICKENING HERD

FEBRUARY CARRIES a quiet but powerful transformation across farmlands and grazing hills.

It is the month when herds begin to stir with new life.

When the udders swell with returning milk and when the rhythm of agricultural households shifts from the stillness of winter to the tender responsibilities of early spring. Before modern calendars and technology, people looked to their animals to understand the turning of the year.

The quickening of herds, especially sheep, signaled that the deepest cold had passed and that nourishment would slowly return. Milk in February held meaning far beyond its practical use. It marked renewal, hope and the gentle warmth that comes from caring for animals at the threshold of a new season. In cultures across Europe and beyond, early milk played a role in rituals, folk beliefs and daily survival. It was one of the first signs that the earth was waking.

Folk Beliefs About Milk in February

Milk has always been more than food. In the late winter months, it became a symbol of life's persistence. When stores of grain and vegetables began to run thin, milk offered nourishment and reassurance. Many households depended on livestock not only for material survival but for spiritual comfort. The first flow of milk in February was often celebrated with gratitude.

> **Milk on the Threshold**
>
> Some households placed a bowl of milk at the doorway on the first days of February. If the milk remained undisturbed overnight, it was taken as a sign that the family would be protected and well provided for in the year ahead.

In Celtic regions, milk was associated with Brigid, guardian of hearth, livestock and healing. Offerings of milk were poured on thresholds or left near stables as a sign of respect for the returning abundance. A bowl of milk left overnight was believed to invite blessing, and if it was untouched by morning, the family took it as a sign of protection.

Elsewhere, farmers believed that early milk could predict the coming season. If the consistency was rich, the year promised prosperity. If thin, the land might struggle. In some communities, mothers shared warm milk with neighbors to strengthen bonds during the lean months. Milk stood for generosity as much as survival.

Stories also warned that ill fortune could follow those who carelessly spilled fresh milk in February. This taboo reflected the preciousness of every drop. Milk was a first fruit of the year, the earliest sign of renewal, and communities treated it with deep respect.

Agricultural Cycles of Lambing

Lambing season begins in late winter for a reason shaped by nature's clever timing. Sheep carry their young through the darkest months and give birth as daylight increases.

This ensures that lambs are born when temperatures begin their slow rise and when spring grass will soon be available.

The arrival of lambs brought a tide of activity to farms that had been quiet since autumn. Shepherds kept long hours, watching barns and fields for signs of labor. Newborn lambs were dried, warmed and encouraged to stand. Their soft bleats filled the cold air with a sound that meant life had returned to the land.

The ewes' milk, rich with nutrients, was essential for the lambs' survival. Its return also signaled that humans, too, would soon have access to milk after months of scarcity. Lambing was therefore both practical and symbolic. It represented the household's ability to endure winter and prepare for the abundance ahead.

Farmers who understood the rhythms of their animals could predict spring's progress with remarkable accuracy. The timing of milk flow, lambing patterns and the behavior of the herd provided cues that the earth itself was shifting. The quickening herd mirrored the quickening of the landscape as sap rose in the trees and early plants broke through frost.

Nutritional and Symbolic Meaning of Early Milk

Early milk is biologically and symbolically unique. In its first days, it is rich with nutrients designed to support fragile newborns. Thick, golden colostrum contains vital proteins and protective compounds. Its purpose is to strengthen and fortify. Communities recognized this richness long before understanding its chemistry. They saw early milk as a source of vitality.

> **Shepherding Notes**
>
> **Colostrum**
> • First milk produced after lambing
> • High in protective proteins
> • Essential for newborn health
>
> **Lambing Patterns**
> • Triggered by increasing daylight
> • Most intense in February
> • Corresponds with early seasonal shifts
>
> **Historical Importance**
> Early milk supported households during late winter scarcity.

In many traditions, early milk was used in ceremonial foods or simple puddings prepared to welcome the shifting season. It marked the return of nourishment after a time of scarcity. Families believed that consuming early milk helped align the body with the returning sunlight and renewed energy of the land.

Symbolically, milk represents purity, nurturing and continuity. It flows when life is present, making it a powerful emblem of renewal in late winter. Milk in February tells a story that grain cannot. It speaks not of storage but of new beginnings.

To drink warm milk during this time connected people to their animals, their land and their ancestors. It affirmed that the coldest part of the year had been endured and that the household would be sustained in the weeks ahead.

Even today, the presence of lambs in fields and the first flow of milk remind us that renewal begins quietly. It begins with nourishment, with care and with the gentle quickening of life returning to the world.

Reflection and Ritual for Milk and the Quickening Herd

If you wish to honor this ancient seasonal moment, take a quiet moment with a warm drink, whether milk, a plant based alternative or any comforting cup. Sit near a window where early light enters. Hold the cup between your hands and feel its warmth.

Reflect on what nourishment means to you at this time in your life. What sustains you? What new beginnings are forming just beneath the surface? Consider how the land prepares slowly, how animals awaken at the right moment and how your own growth might be following a natural cycle.

If you wish, write a single word of blessing or intention on a small piece of paper and place it beneath your cup. Let it rest there for a moment before folding it and keeping it somewhere meaningful.

Just as milk returns to the herd in February, allow yourself to receive what nourishes you gently and fully.

13 / WAX AND FLAME
THE MATERIALS OF RETURNING LIGHT

IN THE DIM stretch of late winter, before the sun has fully reclaimed the sky, people have long turned to candles as symbols of hope, guidance and renewal. A candle flame is small, yet it changes a room. It warms cold corners, softens shadows and makes the quiet hours feel alive.

> *In February, when the world stands between winter's hold and the promise of spring, wax and flame take on a special meaning.*

They embody the fragile but growing light that nature is beginning to restore. The materials used to create candles reveal deep history and intuition. Wax, whether from bees or plants, captures the essence of stored sunlight. Wick fibers hold memory of growth, whether spun from flax, cotton or rushes. Fire transforms these quiet materials into a steady glow that speaks directly to the human heart. Candle rituals, passed down through generations, acknowledge the turning of the year long before scientific calendars named it.

Candles are more than objects. They are companions of the season, bearers of brightness in the year's darkest hours and markers of transition from stillness into movement.

Candle Lore and Household Rites

Long before electricity, the household candle was both practical and sacred. In winter, when days were short and nights stretched long, people relied on candles for light, warmth and spiritual comfort. Their glow united families around hearths and tables. Their flames illuminated rituals meant to protect the home, bless the year ahead or honor the slow return of daylight.

Many households kept a special candle for rites performed at the start of February. This candle was carried through each room to cleanse stagnant energy and invite clarity.

Some families placed candles in windows to guide good fortune toward them or to symbolically welcome returning sunlight. Others lit candles at dawn to strengthen the lengthening day, believing the flame helped encourage the sun itself.

Candle lore varies across cultures, yet a few themes repeat again and again. A bright, steady flame foretold calm and prosperity. A flickering flame suggested emotional or weather shifts. Extinguishing a candle with breath was sometimes discouraged, for it was believed to disturb the spirit of the flame. Instead, many households used damp fingers or a candle snuffer to close the light gently.

In some regions, beeswax candles were considered superior for winter rites. Bees collect sunlight in their work, storing summer energy in honeycomb. When beeswax burns in February, the flame is said to carry the warmth of seasons past into the cold months still ahead.

Traditional Candle Materials

Beeswax
- Bright, clean flame
- Mild honey scent
- Symbol of purity and stored sunlight

Tallow
- Used in practical household candles
- Softer glow
- Represented resourcefulness during winter scarcity

Rushlights
- Made from rushes soaked in fat
- Common among rural families
- Reminders of thrift and seasonal ingenuity

Physics of Flame, Heat and Light

A candle flame may appear simple, yet it is a precise balance of chemistry and physics. Wax melts near the wick and rises through its fibers by capillary action. Once vaporized, the wax burns as a gas, producing heat, light and the steady glow that has accompanied human beings for thousands of years.

The lower part of the flame glows blue, where oxygen feeds the hottest region. The middle shines gold because tiny carbon particles heat until they emit warm colored light. The outer envelope of the flame completes combustion, creating the familiar teardrop shape. Each part works in harmony to maintain stability and brightness.

Heat from the flame creates a small rising column of air that carries warmth upward.

In cold rooms or winter barns, this movement feels gentle but steady, reminding people that even a small flame can shift the atmosphere around it. The physics of light contributes to the comforting feel of candles. Their warm spectrum of color stimulates calm and helps regulate circadian rhythms affected by winter darkness.

Understanding the science does not diminish the magic.

If anything, it deepens appreciation. What feels like a simple flame is actually a balanced conversation between matter and energy. Wax shifts into vapor. Vapor becomes light. Light becomes guidance. The candle reminds us that transformation is both delicate and powerful.

Why Candles Symbolized Hope in Winter's End

Candles carry emotional meaning that reaches far beyond their physical form. In winter, when natural light is scarce, a flame becomes a beacon of intention. It tells the body that warmth is possible. It tells the mind that renewal is coming.

It tells the heart that darkness does not last forever.

For many communities, lighting candles during February symbolized encouragement for the returning sun.

People saw themselves as participants in the changing year, supporting nature's shift through small rituals of brightness. Candles marked the halfway point between winter and spring, a moment when patience felt thin but hope grew stronger.

Candles also symbolize inner light. During late winter, spirits often feel quiet or tired. A small flame offers gentle companionship. It invites reflection and steadiness. It reminds people that growth happens slowly at first, just as the lengthening daylight begins not with sudden brightness but with a few extra moments each morning.

In a world where technology offers abundant light, the candle still touches something ancient and universal. We lean toward it instinctively. We watch it. We breathe with it.

We sense the promise held within its small, steady glow.

Reflection and Ritual for Wax and Flame

Choose a candle that feels meaningful to you. It may be beeswax, tallow, soy or any material that carries a warm, gentle scent. Sit in a dim room or near a window in early evening.

Light the candle and watch its flame settle. Notice its colors and the way it moves with subtle rhythms. Allow your breath to match its calm rise and fall.

Reflect on what you are welcoming into your life as winter draws to its close. What inner light needs tending? What quiet hope is beginning to take shape?

If you feel called, speak a simple intention while touching the candle's base. Then let the flame burn for a few minutes before gently extinguishing it with care.

Carry the calm brightness with you into the lengthening days.

14 / ASH AND EMBER
THE MEMORY OF THE WINTER HEARTH

A WINTER HEARTH holds stories long after its flames have faded. In February, when fires still warm the evenings but the world begins its slow turn toward spring, the residue of burning wood becomes a symbol of memory, endurance and renewal. Ash and ember are not simply remnants of fire. They are records of warmth, guardians of the home and materials woven deeply into folk tradition.

Where flames dance brightly for a brief moment, ash and ember endure. Ash settles softly, pale as winter dust. Embers glow long after the blaze has subsided, holding heat that lingers like a quiet heartbeat beneath the surface. These final forms of fire mirror the seasonal moment of February itself. The brightest heat of winter has passed, yet its strength remains, ready to nourish what comes next.

Communities across the world recognized the power hidden in ash and ember. They used them in household rites, agricultural practices and symbolic rituals that honored both the fire and the land.

Ash Charms and Purification Traditions

Ash has been part of cleansing ceremonies for centuries. After winter fires burned through long nights and protected families from cold, their ash became a material of gratitude and renewal. Many cultures believed that ash carried the memory of flame and that its soft gray texture held protective qualities.

In rural households, a small amount of ash was sometimes placed on thresholds or around livestock pens to guard against misfortune. Farmers scattered ash in fields to bless the soil or to mark the boundaries of newly tilled land. Families drew simple shapes in ash during early February rites to invite health, clarity or emotional grounding.

Some traditions linked ash to ancestors. As fire was seen as a messenger between worlds, its ash was viewed as a connection to those who had passed. A thin layer of ash sprinkled near a hearth stone was believed to honor familial spirits and invite their guidance for the coming season.

In religious contexts, ash often symbolized humility and remembrance. It marked a turning point from winter reflection toward spring renewal. The act of placing ash upon oneself or within the home made tangible the cycle of letting go and preparing for growth.

. . .

Ash Line on the Threshold

In several regions of Europe, families drew a thin line of ash across the doorway after a winter storm to protect the household and to signal that the season was shifting into gentler days.

Nutrient Cycles in Wood Ash

Ash holds value far beyond symbolism.

> *Scientifically, wood ash is rich in minerals that return nutrients to the land.*

When trees burn, the organic material disappears into heat and light, leaving behind concentrated minerals that once helped the tree grow.

Wood ash contains calcium, potassium, magnesium and trace elements that support soil health. Farmers recognized its benefits long before understanding its chemistry. They spread ash across gardens, orchards and fields to enrich the soil at a time when the ground was preparing for new life.

Traditional Hearth Ash Uses
• Protection charms placed at doorways
• Purification lines drawn near livestock pens
• Winter soap making using ash lye
• Soil enrichment for gardens preparing for spring
• Symbolic boundaries for rituals of renewal

The act of scattering ash in February symbolized feeding the earth with the memory of past seasons.

. . .

> **Properties of Wood Ash**
>
> **Calcium**
> • Strengthens plant structure
> • Moderates soil acidity
> **Potassium**
> • Supports early spring growth
> • Improves water regulation in plants
> **Magnesium and Trace Minerals**
> • Aid in chlorophyll production
> • Help balance soil nutrition
> **Practical Uses**
> • Natural soil amendment
> • Moisture absorber
> • Ingredient in historical soap making

The minerals in ash raise soil pH, making it especially useful for acidic ground. The presence of potassium supports root development and early growth, making ash a natural ally in agricultural cycles. Even when mixed lightly into compost piles, ash encourages faster decomposition and balances moisture.

In this way, ash embodies transformation. Wood becomes fire. Fire becomes ash. Ash becomes nourishment for future growth. The cycle reflects the transition from winter into spring, where the memory of what has been supports the promise of what will form.

Embers as Symbols of Endurance

Embers are the quiet core of winter fire. Even after flames vanish, embers glow with steady heat, holding the last warmth of a long night. Their persistence made them powerful symbols of endurance, patience and quiet strength.

In many cultures, the morning hearth was rekindled from a bed of embers rather than from fresh flame. This practice not only conserved resources but symbolized continuity. The home's fire never fully died. It rested, waiting for the next moment to rise.

In February, embers represented the inner fire needed to carry communities through winter's final stretch. While the world outside remained cold, embers reminded people that warmth still lived beneath the surface. The ember's glow offered comfort during early morning chores and late evening reflection.

. . .

Symbolically, embers represent the enduring spark within each person. When life feels stalled or dim, the ember suggests that energy remains, ready to be nurtured. A single breath or gentle stirring can bring renewed brightness. This imagery resonates deeply with the season.

> *February is the ember of the year, holding quiet heat until spring ignites.*

Some households believed that blowing gently on an ember brought blessings for the day, while tossing herbs or dried flowers into glowing coals carried prayers toward renewal. Embers carried intention, warmth and memory.

Reflection and Ritual for Ash and Ember

Gather a small amount of cooled wood ash, or use a symbolic pinch of soil if ash is unavailable. Place it on a simple cloth or bowl. Sit in a quiet space beside a candle or warm light.

Touch the ash lightly with a fingertip. Consider what memories or lessons winter has offered you. What has burned away? What remains useful? What can now nourish your next steps?

If you wish, draw a small shape or symbol in the ash. Let it represent clarity, protection or renewal. When finished, release the ash outdoors to the wind or scatter it gently into soil.

Then look into a candle flame and imagine the ember behind it. Trust that even small sources of warmth and light hold enough strength to guide you into the coming season.

PART 3
SKY, EARTH AND THE SCIENCE BEHIND THE STORIES

15 / WHY FEBRUARY IS A TURNING POINT

February sits at a quiet threshold in the natural year. It is not the radiant awakening of spring nor the deep stillness of winter, yet it carries the seeds of both. Across cultures, traditions developed around this moment because people sensed something important happening in the world around them.

> *Animals behaved differently. Light felt subtly stronger. People's moods shifted. Snow melted and refroze in quick succession. Something in the air changed, even if it could not be named.*

Today, science reveals the mechanisms behind these ancient observations. February is a turning point because the planet, the atmosphere, plant life, wildlife and human biology begin to react to increasing light long before temperatures catch up. Folklore noticed these truths through lived experience. Modern science explains them through measurable patterns. Together they tell the same story: February is when the natural year begins to stir.

The Planet's Tilt

The reason February feels different lies far beyond Earth itself. Our planet orbits the sun while tilting at an angle of about twenty three degrees. This tilt creates the seasons and determines how sunlight reaches the surface at various times of year.

> *During winter in the Northern Hemisphere, the North Pole leans away from the sun. Sunlight arrives at a low angle, spreading thinly across the land and creating short, dim days.*

By February, the Earth has moved far enough along its orbit that the Northern Hemisphere begins to tilt closer to the sun again. This shift increases the height of the sun in the sky, even though most people still feel the cold.

This change is subtle but important. The sun rises earlier, sets later and follows a slightly higher arc each day.

> *Even a small increase in solar angle delivers more energy to the surface.*

That then begins to influence soil temperature, plant chemistry and the behavior of animals. Ancient communities, lacking modern tools, observed the consequences of this tilt through nature. They saw catkins lengthen, birds call more loudly and certain seeds begin to swell beneath frost. The light had changed. The year had shifted. The planet's tilt had set everything in motion.

Increasing Light

February's most meaningful transformation is not warming air but growing daylight. Light increases more rapidly now than at any time since the summer sun began its long descent.

In many regions, each day brings one or two additional minutes of brightness at dawn and dusk. This small but steady expansion creates powerful biological effects.Plants begin responding weeks before they visibly bloom.

Day length influences hormones inside buds.

As light increases, these hormones trigger cell growth, prepare leaves for emergence and signal roots to take up more nutrients. Many of the plants celebrated in folklore for appearing early, such as snowdrops and willow, respond primarily to light rather than temperature. Their internal clocks are tuned to the sky.

Animals also respond to February's light. Birds begin practicing their mating calls, not because the air is warm but because light influences hormones that prepare them for reproduction.

Mammals become more active. Insects begin stirring in sheltered places on sunnier days. Even beneath snow, soil microbes wake and begin their slow seasonal work.

Light is the true engine of seasonal change. February holds a sense of renewal because the increasing brightness affects every level of the natural world. The air may still feel cold, but the rhythm of life is already shifting.

. . .

Effects on Human Mood and Wildlife

Humans feel February's turning point as clearly as plants and animals. Our bodies respond to light just as theirs do. More daylight reduces melatonin levels, which encourages alertness in the morning and improves energy throughout the day.

Serotonin and dopamine patterns begin to shift as well, gently lifting mood.

People often describe February as a time of new motivation or emotional awakening, even if they cannot explain why.

For many, the emotional weight of deep winter begins to soften. The mind feels clearer. The heart begins to open to new ideas or projects.

This psychological shift aligns beautifully with the seasonal rituals described in folklore, which focus on renewal, purification and preparation for spring.

Wildlife also views February as a time of readiness. Birds begin staking out territories. Foxes and owls engage in courtship. Small mammals emerge more frequently in search of food. Many species depend on light to synchronize breeding cycles so that young are born when food becomes available.

. . .

Even aquatic life responds. Longer daylight affects algae growth beneath ice and influences the movements of fish. February's changes ripple through every ecosystem, weaving together behaviors that ensure survival and future abundance.

What appears quiet on the surface is actually a symphony of awakening. February is less a pause between seasons and more the first quiet breath of the new year.

Reflection and Ritual for the Turning Point

Step outside on a February morning, even briefly. Notice how the light touches the ground, the angle of shadows and the faint warmth carried by the sun. Whether the day is cold or mild, the growing light is present.

Consider the subtle shifts happening around you in the natural world. Buds tightening in readiness. Birds watching from bare branches. Soil waking beneath frost. Let these small signs remind you that change often begins quietly, long before it becomes visible.

If you wish, write a sentence describing what the returning light awakens in you. Fold it and keep it in a safe place until spring, when you can revisit it under brighter skies.

16 / THE HUNGER MOON
OLD NAMES FOR HARSH DAYS

FEBRUARY's full moon has carried many names across cultures, but one of the oldest and most evocative is the *Hunger Moon*.

> *This name arose during a time when winter stores ran thin, game was scarce and the coldest nights still lingered even as the year began to shift.*

The Hunger Moon illuminated a landscape caught between endurance and renewal. Its light fell on frozen fields, quiet forests and communities that relied on careful rationing to survive the final stretch of winter.

The moon's glow was both beautiful and stark. It reflected the truth that the natural world, too, faced limits during late winter. Animals moved cautiously. Plants remained dormant, waiting for warmth. People measured their grain stores and hoped for a steady return of forage, fresh milk and early greens. The Hunger Moon did not symbolize despair so much as realism, awareness and resilience.

February invites honesty about the tension between scarcity and renewal. The old name reminds us that endurance carries wisdom, and that nature's rhythms build strength in the quietest, leanest times.

Folklore of Scarcity

Communities once shaped their lives around the cycles of abundance and scarcity. The late winter period, illuminated by the Hunger Moon, brought a heightened awareness of limits. Deep winter stores of cured meat, root vegetables and grains were dwindling. Families learned to stretch supplies, practicing thrift and gratitude. In the stories told during this time, hunger became a teacher.

Folklore across Europe and North America speaks of spirits or guardians who roamed during the Hunger Moon, reminding people to be humble, share what they had and avoid waste. Some tales warned that greed during these harsh days might invite misfortune.  Others described magical creatures emerging at night, testing a traveler's kindness with requests for food or warmth. These stories guided people toward generosity, even when they had little to give.

> *People also believed that dreams occurring under the Hunger Moon held messages about the coming spring.*

A dream of full baskets or returning livestock was interpreted as a sign of abundance ahead. A dream of frost or barren fields encouraged caution. These interpretations helped families cope with uncertainty by offering psychological comfort and a sense of preparedness.

The name Hunger Moon preserved the cultural memory of endurance. It honored the reality of lean times while carrying a quiet promise that winter's end was near.

. . .

Energetics of Late Winter Ecosystems

Nature, too, enters a state of scarcity in February. While light is increasing, temperatures remain low, and food availability for wildlife remains limited. This creates a dynamic period in which ecosystems stretch their resources with remarkable efficiency.

Plants, still dormant, hold energy deep within roots.

Photosynthesis begins to increase as daylight grows, but visible growth remains halted. The soil is cold and slow. Fungi and microorganisms awaken cautiously, beginning their slow work of breaking down organic matter. These subtle processes prepare the land for spring but provide little active food for animals.

Predators face significant challenges in late winter. Snow cover can make prey harder to scent or track, yet prey animals become weakened by months of limited forage. Energy budgets tighten throughout the ecosystem.

Every movement, every decision and every search for shelter or food must be measured. Yet this measured pace allows ecosystems to conserve energy until warmth unlocks more abundant resources.

. . .

February landscapes hold a strange tension. Stillness dominates, but beneath the surface, the first exchanges of renewed life have already begun. The Hunger Moon shines on an ecosystem in waiting, poised for the eruption of growth that will come with spring.

How Animals Ration Resources

Animals have evolved finely tuned strategies for surviving the scarcity of late winter. Some slow their metabolism. Others alter daily patterns or reduce unnecessary travel. Many ration energy rather than food itself, adjusting behavior to conserve strength.

Deer limit movement, staying within familiar wintering areas where shelter and minimal forage are reliably found. They conserve calories by choosing sheltered resting places and reducing long treks across snow.

Birds alter feeding patterns to maximize the brief daylight hours. Some species practice micro fasting overnight, allowing body temperature to drop slightly to conserve energy. Others gather in communal roosts to share warmth.

Small mammals such as squirrels and mice ration stored nuts or seeds, often using complex memory maps to locate hidden caches. During February, these caches become critical lifelines.

In harsh winters, animals sometimes leave food untouched for days to maintain reserves for even tougher conditions.

Predators also ration through behavior. Foxes conserve energy by listening for prey under snow rather than patrolling wide territories. Owls hunt from still perches, minimizing physical exertion. Every action in late winter balances risk and reward. Even aquatic animals adjust. Fish slow their movement in cold water to conserve energy. Amphibians remain in hibernation states, surviving on stored fat until temperatures rise.

. . .

These instinctive strategies reveal the intelligence of nature's cycles. Animals treat February with respect. They understand its challenges and its significance. They move through scarcity with patience, preparing quietly for the abundance that will soon return.

Reflection and Ritual for the Hunger Moon

On the night of the Hunger Moon, step outside or look through a window at the full moon's light. Notice how clearly it illuminates the landscape, even in the deep cold of winter.

Reflect on the lessons of scarcity and endurance. Ask yourself where you might conserve your energy or which habits you can release to prepare for renewal. Consider what resources you hold within yourself that will carry you through your own transitions.

If you wish, write a brief intention about caring for your inner reserves. Fold the paper and place it under a stone, symbolizing patience and steadiness. Return to it when spring arrives and notice how your perspective has changed.

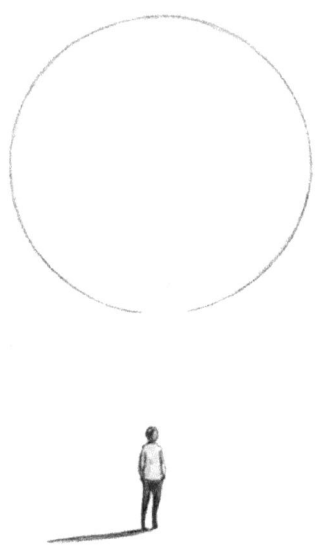

17 / READING THE SKY
READING THE SKY

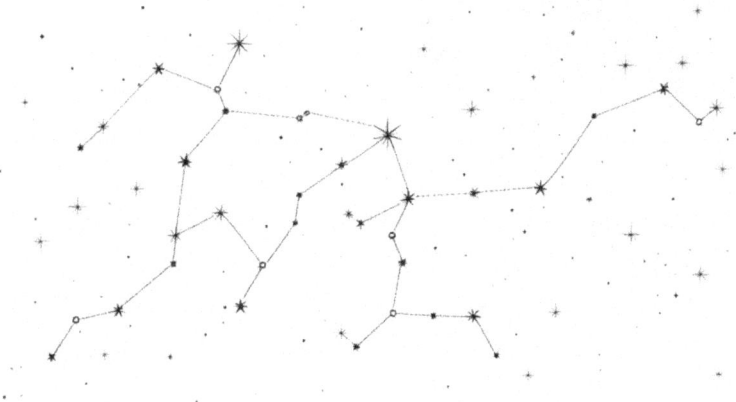

For most of human history, people read the sky as both map and messenger.

The heavens offered direction for travelers, inspiration for storytellers and guidance for farmers waiting for the right moment to sow or harvest.

February's sky is clear and sharp, its stars bright against cold air that steadies light and sound. When days begin to lengthen but nights remain long, the sky becomes a canvas rich with constellations, shifting weather signs and whispers of the season's turning.

. . .

Communities once watched February's sky closely. Constellations served as seasonal anchors, appearing each year in familiar patterns.

Weather rhymes arose from generations of careful observation.

And long before meteorology, people sensed atmospheric changes by reading winds, clouds, color and the behavior of animals. Some folklore holds surprising accuracy. Other tales reveal insight shaped by landscape and survival.

February's sky is honest. It reveals both the challenges and hopes of late winter. It teaches patience, preparation and trust in the larger cycles that shape the year.

February Constellations

February nights are dominated by some of the brightest and most recognizable constellations in the northern sky. The clarity of winter air makes them appear near and vivid, like lanterns suspended in cold darkness.

Orion

Orion strides high in February, his belt forming a line that points toward Sirius, the brightest star visible from Earth. Orion's presence in folklore is widespread. Hunters, warriors and guardians take shape in his stars. His appearance signals the depth of winter but also the approach of seasonal change.

Canis Major and Sirius

Following Orion is his great dog, marked by Sirius. In many traditions, Sirius was a weather star. Its brilliance was said to foretell clear nights or sudden frosts. Although not a reliable predictor, its presence near Orion helped ancient observers track seasonal progress with surprising precision.

Taurus and the Pleiades

Taurus and the small cluster of the Pleiades sit west of Orion. Farmers once believed the Pleiades' shimmering tight cluster could reveal weather conditions. A crisp, clearly visible cluster suggested cold, dry air. A blurred cluster hinted at moisture and approaching snow.

Auriga and Capella

Auriga, the charioteer, sits overhead with Capella, a bright star often associated with storm warnings in folklore. Capella flickers noticeably during turbulent winter nights because of atmospheric motion, leading communities to interpret its dancing glow as a sign of unsettled weather.

These constellations return each February, offering a sense of continuity. They remind us that the earth's slow turning and the sky's steady rhythms guide more than night navigation. They carry memory, pattern and story.

What Weather Rhymes Get Right

Weather rhymes and sky sayings were tools for survival. They condensed generations of observation into short, memorable phrases. Although some are charming superstitions, others contain kernels of scientific truth.

"Clear moon, frost soon."

This holds accuracy in many climates. A clear sky allows Earth's heat to escape into the upper atmosphere. Without cloud cover, the ground cools rapidly at night, making frost more likely.

"Red sky at night, sailor's delight. Red sky in morning, sailor take warning."

This rhyme reflects principles of weather systems. A red sunset often indicates dry, stable air moving in from the west. A red sunrise can result from incoming moisture or storms illuminated from below by early light.

"Halo around the moon, rain or snow soon."

A halo typically forms when moonlight passes through high, thin cirrus clouds made of ice crystals.

These clouds often appear before a storm front, making precipitation more likely within a day or two.

"If the groundhog sees his shadow, winter will persist."

This lore is less accurate, but it reflects a broader truth: bright, clear days in early February can coincide with lingering cold, while cloudy conditions often accompany warmer air masses.

Weather rhymes endure because they offered guidance, not certainty. They captured the essence of February's unpredictability while providing comfort and structure.

Atmospheric Patterns of the Season

February's atmosphere holds tension between winter's grip and spring's approach. This creates patterns that shape the sky, weather and everyday experience.

Stable, Cold Nights

Cold air is dense and sinks, especially under clear skies. This stability creates crisp, star filled nights when sound carries far and frost forms quickly. The clarity makes February ideal for stargazing.

Changing Jet Stream Behavior

The jet stream begins to shift in late winter. Its position influences storm paths, temperature swings and wind patterns. Some years it dips low, allowing arctic air to descend. Other years it lifts, inviting milder weather.

Intermittent Thaw and Freeze

The increasing sun angle brings enough warmth to thaw the upper layer of snow during the day. Nights refreeze it into a hard crust. Animals navigate these shifts with care, and plants respond slowly beneath the soil.

Atmospheric Scattering of Light

With the sun rising slightly higher each day, morning and evening skies take on new colors. February often brings pearlescent dawns and vivid sunsets due to fine winter particulates scattering light.

Increased Winds

As temperature contrasts shift between land and sky, winds become more active. These winds shape clouds, influence bird behavior and contribute to the changeable mood of late winter weather.

Atmospherically, February is neither stagnant nor settled. It is a month of balancing forces, of light overcoming dark, of cold negotiating with warmth. The sky reflects this negotiation with beauty and unpredictability.

Reflection and Ritual for Reading the Sky

Go outside on a clear February night. Look for Orion's belt, the shimmer of the Pleiades or the steady glow of Sirius. Let the cold air sharpen your senses.

Notice how the sky feels timeless yet alive. How the stars seem steady while the atmosphere shifts. Imagine your ancestors reading these same constellations as guides for survival and hope.

If you wish, choose one star and watch it for a moment. Let it anchor your thoughts. Breathe slowly and consider what patterns in your own life are becoming clearer as the season begins to turn.

18 / SONGS IN THE COLD

SONGS IN THE COLD

February may appear silent at first glance, yet the air carries a subtle shift. Before leaves return, before insects rise, before the soil fully warms, birds begin to sing. Their voices weave through bare branches and frozen mornings, announcing the earliest stirrings of courtship long before spring is visible.

These songs feel brave in the cold, as though the birds know something the season has not yet revealed.

People across cultures noticed this late winter chorus and wove stories about love, pairing and renewal around it. Birdsong in February symbolizes hope and readiness. It signals that life is preparing to unfold again. Even when frost lingers, the notes of blackbirds, finches, chickadees and doves create a soundscape that softens winter's edge.

This chapter explores why birds sing now, the folklore surrounding avian courtship and the unique soundworld of late winter.

Why Birds Start Singing Now

Birdsong at this time of year is tied to increasing daylight rather than warmth. As the days lengthen, light enters birds' eyes and stimulates hormonal changes that signal the beginning of breeding season. Even if snow still carpets the ground, light awakens their internal rhythm.

Male birds begin practicing early versions of their courtship songs. These are quieter and less elaborate than their full spring performances, almost like warm up melodies. These early songs serve several purposes:

Territory Claims

Birds begin marking boundaries before competition intensifies. Singing in February makes a location known and discourages rivals.

Mate Attraction

Although many females are not ready to choose mates yet, early singing helps males refine their calls. Strong, steady vocals demonstrate health and vitality.

Pair Bond Reinforcement

Some species, such as doves or certain songbirds, maintain long term partnerships. Late winter duets help couples reaffirm their bond.

This early singing is a delicate negotiation between caution and preparation. Birds expend precious energy to create these sounds, yet the investment supports future success. Light fuels the signal. Song carries the message. The season begins to shift.

Folklore of Avian Love and Pairing

Birds became powerful symbols of love because their courtship begins while the world still lies in winter quiet. Many traditions believed that birds chose mates in February, especially around the middle of the month. This idea shaped early European customs and influenced later cultural developments surrounding human affection.

In medieval England, people once said that the fifteenth of February was the day birds selected their partners. Lovers offered one another small tokens inspired by the birds' example. Rural communities watched for courting pairs of doves, finches or blackbirds as signs that the year of growth had begun.

In Scotland, seeing two small birds perched close together on an early February morning was considered a blessing for household harmony. In some regions, young people performed playful divination, drawing bird names from a basket to foretell the character of a future partner.

Across cultures, specific birds carried symbolic meaning:

Doves

Known for gentle nature and steady partnerships, doves represented loyalty and peace.

Finches

Bright and lively, they symbolized joy and youthful affection.

Blackbirds

Their bold songs cut through winter silence. In folklore, they represented strong hearts and resilience in love.

Owls

Although not always symbols of romance, owls were said to watch over nighttime courtships and lend wisdom to those seeking connection.

These stories reveal how deeply people felt attuned to the natural world. Birds sang, and humans listened. The sound of pairing in cold air inspired hope and tenderness long before flowers or warmth returned.

Soundscapes of Late Winter

Late winter sound carries a particular texture. Cold air shapes how sound travels, often making it sharper and more resonant. Birdsong in February rings clearly across quiet landscapes because competing noises are scarce.

> *The absence of leaves allows notes to travel farther, and frost covered ground reflects sound with a crisp brightness.*

Different bird species contribute distinct voices to this seasonal soundscape:

Chickadees

Their clear, whistled calls brighten morning air. They often adjust their song structure in late winter, giving hints of approaching courtship.

Blackbirds and Thrushes

These early singers offer melodic runs that sound almost out of season.

Their voices emerge confidently from hedgerows or treetops.

Finches

Finches add rapid trills and cheerful phrases, creating lively patterns even on cold days.

Doves

Soft coos drift gently across fields, giving the soundscape a soothing undercurrent.

Beyond birds, the winter world contributes its own music.

Snow creaks underfoot. Bare trees tap lightly in the wind. Ice cracks on ponds as temperature shifts. Together these sounds create a quiet orchestra tuned to the season's first signs of renewal.

Birdsong in February reminds us that new life begins with voice before it begins with color.

Sound awakens the world long before buds open. Early song is an act of trust. It reaches into cold air and believes in the warmth that is coming.

Reflection and Ritual for Songs in the Cold

Wake early on a February morning and step outdoors or open a window. Listen for the first calls carried through the quiet air.

Ask yourself which song resonates with you. Is it the confident melody of a blackbird, the hopeful trill of a finch, or the soft reassurance of a dove?

Close your eyes and let the sound settle into your breath. Allow the early chorus to inspire a sense of readiness. Consider what beginnings you are preparing for, what intentions you are quietly shaping and what version of your own voice you are ready to offer the coming season.

Write a single sentence describing what late winter birdsong awakens in you. Keep it somewhere you will see again in early spring.

19 / STIRRINGS
   BENEATH THE SOIL

To the eye, February soil appears still. Frozen ground, muted colors and quiet fields suggest a landscape in deep sleep. Yet beneath the surface, life is far from dormant.

> *Roots sense the shift in daylight long before stems or leaves*
> *reveal any change.*

Sap begins its slow climb, carrying the first sweetness of the year. Microbial communities stir, stretching awake from winter's long cold pause.

This hidden world, where darkness meets slow warmth, is the true beginning of spring. Folklore once spoke of sleeping earth spirits who breathed softly underground at this time of year, preparing to open the soil from within. Modern science echoes that ancient intuition, revealing the delicate processes that prepare the land for renewal.

February's surface may seem silent, but the soil below hums with life returning.

Roots Waking and Sap Rising

Even while frost still tightens the upper crust of the earth, roots begin subtle changes that signal readiness for growth.

These early stirrings do not depend on warmth alone. Increasing daylight triggers hormonal shifts within trees and perennial plants.

Photoreceptors in dormant buds sense the growing length of day and send chemical signals downward into the roots.

Roots respond in several ways:

Moisture Uptake Begins

They start drawing water more actively, preparing internal pathways for spring growth.

Nutrient Mobilization

Stored nutrients held within roots begin to loosen and move upward, supporting the earliest metabolic activity in branches and buds.

Sap Flow Strengthens

In many trees, especially maples, sap begins to rise with surprising force.

Cold nights and slightly warmer days create pressure differences that push sap upward through the trunk.

This is why sap collection traditionally begins in late winter, when the tree's first energetic awakening is underway.

To the plant, these processes are not optional.

They are the slow opening notes of the biological song that will lead to spring flowering, leaf growth and fruiting. Roots wake first because they must prepare the whole organism for the season ahead.

Myths of Sleeping Earth Spirits

Long before the science of root pressure or soil chemistry, people understood that something powerful moved beneath the ground in late winter.

> *Folk stories describe earth spirits stretching in their slumber, breathing warmth into frozen soil. Some traditions pictured them as gentle giants turning over within the earth, causing thaws.*

Others imagined small guardians who tended underground chambers where seeds rested, waiting for the moment to rise.

Across Europe, Asia and the Arctic, stories shared common themes:

Slumbering Earth Mothers

Many cultures envisioned the land itself as a maternal figure who slept through winter and stirred slowly in February. Her breath was said to soften the soil. Her heartbeat guided roots upward.

Root Keepers and Seed Spirits

Some tales described tiny beings who protected roots from frost. They guarded underground pathways and coaxed early shoots toward the surface with lantern lights or whispered guidance.

Dragons or Serpents of the Deep Earth

In other traditions, serpents slept in the soil through winter and began to awaken near early spring. Their stirrings were linked to the thawing of the land and the return of water.

> *These myths honored the mystery of unseen change. People sensed that life returned from below long before it appeared above.*

The stories offered comfort during a season of scarcity, reminding communities that the earth was alive and preparing.

. . .

Although science now explains these processes differently, the poetry of the myths remains true. Something quiet and powerful does move beneath the soil.

Microbiome Activity in Cold Soil

The soil microbiome, a community of bacteria, fungi, protozoa and microscopic organisms, responds to February's changing conditions with remarkable sensitivity. Even when temperatures remain low, these organisms begin shifting from winter dormancy into early activity.

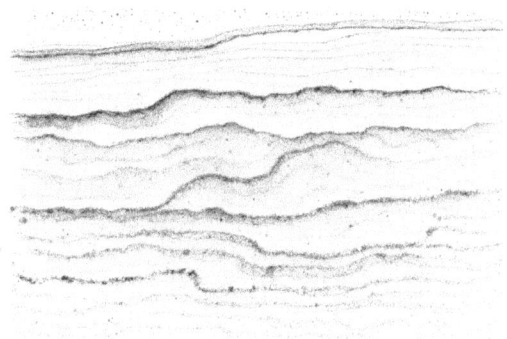

Cold Adapted Microbes Awaken

Some bacterial species function well at near freezing temperatures. As daylight increases, they begin breaking down organic matter again, releasing nutrients that plants will soon absorb.

Mycorrhizal Fungi Reconnect

Fungal networks, which join roots in symbiotic partnerships, begin to reactivate. These networks help plants access nutrients and water, strengthening plants before visible growth begins.

Nutrient Cycling Accelerates

Processes like nitrogen mineralization increase slowly as microbes wake.

Although small in scale, these changes set the stage for spring fertility.

Soil Structure Softens

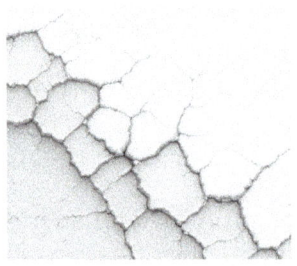

Repeated freeze thaw cycles loosen soil particles. Microbial activity contributes to this softening, creating pockets where roots can expand more easily when warmth returns.

This intricate world beneath the soil is foundational to the health of ecosystems. February's subtle microbial awakening is a reminder that spring begins microscopically before it does macrocosmically.

The earth prepares from within.

Reflection and Ritual for Stirrings Beneath the Soil

Walk outside and rest your hand on the ground, even if it is cold or covered in frost. Imagine the quiet movement below your fingers. Roots gathering strength. Sap rising in silence. Microbes waking in tiny bursts of activity.

Reflect on what is waking within you.

What ideas or hopes lie just beneath the surface? What energy is gathering slowly, preparing you for the coming season?

If you wish, bury a small written intention in the soil or place it under a stone. Let it rest there until spring, trusting that unseen processes will shape both the land and your inner life.

PART 4
LIVING THE MONTH: PRACTICES FOR MODERN READERS

20 / SIMPLE RITUALS OF LIGHT AND PURIFICATION

FEBRUARY CARRIES A QUIET INVITATION. The month sits between deep winter and the early promise of spring, making it a natural time for reset and renewal. The stories, plants, and traditions explored throughout this book point to a shared message: this part of the year is meant for gentle clearing, not dramatic change. It asks for small acts that support clarity, warmth and intention.

Modern life often moves quickly, yet even simple rituals can help reconnect us to the season. You do not need special tools or elaborate steps. You only need a few quiet moments and a willingness to pause. The practices below are designed for contemporary homes and full schedules. They translate ancient ideas into everyday actions that nurture well being.

. . .

Candle Moments

Candles have long been symbols of hope and encouragement in dark months, and their power remains strong today. A single flame can shift the energy of a room and anchor your attention in the present.

How to try this now

- Light a candle during a morning routine while you drink tea or coffee. Let it represent the return of light.

- In the evening, turn off overhead lights for a few minutes and sit with a candle. Notice how the warm glow slows your breath.
- Choose a candle with a scent that makes you feel grounded. Rosemary, pine, or mild floral notes connect naturally with late winter themes.

Purpose

These small candle moments create pockets of calm. They remind you that even the smallest light can brighten a heavy day and support emotional clarity.

Home Cleansing in a Modern Way

Traditional winter cleansing rituals were practical as well as symbolic.

> *Today, many people feel uplifted by clearing physical space at this time of year.*

There is no need to rely on old forms of smoke cleansing or rituals if you do not resonate with. The aim is simple: create space for fresh energy as the season shifts.

How to try this now
• Open a window for a brief moment, even if the air is cold. Fresh winter air refreshes a room quickly.

• Wipe down one surface that matters to you, such as a desk, kitchen counter or bedside table.

• Remove one small item from a cluttered area. This is not a full decluttering project. It is a symbolic gesture that acknowledges transition.

• Add a small natural element, such as a branch, stone or simple houseplant, to remind you of life stirring outdoors.

Purpose
Home cleansing in February is not about perfect order.

It is about choosing simplicity and making room for what feels important as the new season approaches.

Gentle Intention Setting

Heavy resolutions often fail because they clash with the energy of late winter.

February is better suited to quiet introspection and slow shaping of goals.

This is the month to set gentle intentions rather than ambitious plans.

Think of it as planting an early seed rather than expecting full growth.

How to try this now

- Choose one small intention for the month. This could be rest, clarity, kindness or focus.
- Write a short phrase that reflects what you want to nurture. Keep it simple enough to remember throughout the day.
- Place the phrase somewhere visible: on a bedside table, near a desk, or tucked into a journal.
- Spend a moment each week asking yourself how the intention is unfolding.

Purpose

Gentle intention setting helps you align with February's natural energy. You are not forcing change. You are preparing the inner ground for what will grow more easily in spring.

A Seasonal Reset for the Modern Reader

You do not need to practice old traditions exactly as they were.

What matters is the spirit behind them: renewal, care, light and clarity. February offers a moment to pause, breathe and prepare for the upward movement of spring.

Light a candle. Open a window. Clear a corner. Speak a simple intention.

These small acts can bring peace to your home and steady your mind.

The month invites you to notice stirring life both inside and around you.

Let your rituals be easy. Let your pace be humane. Let your sense of possibility grow in time with the returning light.

RITUAL CARDS FOR MODERN READERS

**Ritual Card 1

Morning Light Pause**

Practice: Light a candle for the first minute of your day.

Purpose: Welcome the returning light.

**Ritual Card 2

Window Refresh**

Practice: Open a window for ten breaths, even in cold weather.

Purpose: Invite clarity into your space.

**Ritual Card 3

Clear One Surface**

Practice: Wipe or tidy a single area you use often.

Purpose: Make room for calm focus.

**Ritual Card 4

Small Natural Offering**

Practice: Place a stone, leaf or branch somewhere visible.

Purpose: Connect home energy with seasonal rhythms.

**Ritual Card 5

Simple Gratitude Flame**

Practice: Light a candle and name one thing you are grateful for.

Purpose: Strengthen emotional steadiness.

**Ritual Card 6

Gentle Intention**

Practice: Write one short phrase that reflects your aim for the month.

Purpose: Support slow, steady personal growth.

Ritual Card 7

Five Minute Reset**

Practice: Sit quietly for five minutes, noticing your breath and posture.

Purpose: Reboot your internal pace.

Ritual Card 8

Release One Item**

Practice: Remove one object from a cluttered space.

Purpose: Symbolize letting go of old energy.

Ritual Card 9

Warm Hands Ritual**

Practice: Hold a warm mug with both hands and breathe mindfully.

Purpose: Ground yourself gently.

Ritual Card 10

Evening Candle Closure**

Practice: Light a candle as the day ends. Sit with it for one slow minute.

Purpose: Mark the transition from activity to rest.

Ritual Card 11

Scent of Renewal**

Practice: Use a mild natural scent like rosemary, pine or citrus.

Purpose: Refresh the senses and uplift mood.

Ritual Card 12

Seasonal Journal Seed**

Practice: Write one sentence about what you hope to grow this year.

Purpose: Plant a symbolic seed of intention.

21 / NOTICING THE FIRST SIGNS OF SPRING

Early spring does not arrive in a single moment. It begins in subtle shifts that often go unnoticed unless we pause and look closely.

February and early March contain dozens of small changes that signal the year turning.

These changes may feel quiet, especially if the weather is still cold, but they are real and visible once you know where to look.

Noticing early signs of spring is a practice in slowing down.

It invites you to engage with the world around you in a gentle, observational way. These moments help build a sense of connection to place and season, and they remind you that renewal begins gradually and patiently.- Modern life rarely encourages us to pay attention to micro seasons or tiny details, yet these are the clues that reveal the world waking up.

Micro Seasons

Many cultures break the year into smaller segments than the four familiar seasons. These micro seasons reflect short windows of change that last only a few days. Although often tied to older traditions, they translate beautifully into contemporary awareness practices.

In February and early March, micro seasons may include:

- first birdsong before sunrise
- ice melting in thin sheets rather than solid blocks
- sap rising in trees
- soil softening after frost
- the appearance of the earliest shoots

Noticing micro seasons does not require formal knowledge. It begins with asking a simple question: What is different today from last week?

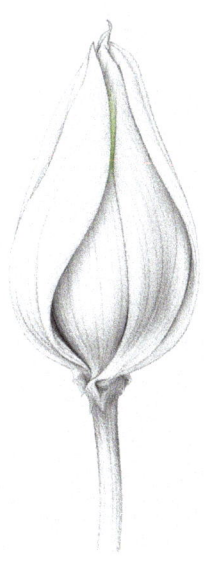

This practice creates a rhythm of gentle attention that helps anchor you in natural time rather than digital time.

Buds and Early Growth

Even when branches still look bare, buds begin to swell. If you look closely, you may see:

- tighter, rounder shapes along twigs
- a slight shift in color from gray to green or brown
- softer outlines as protective bud scales loosen

Trees and shrubs prepare long before leaves actually emerge.

. . .

Once you know what to look for, you will see the entire landscape entering a phase of quiet preparation.

You might also notice early shoots from plants like snowdrops, crocus or hellebore. They often appear while frost still lingers, reminding us that growth does not wait for perfect conditions.

Try choosing one plant or tree near your home and observe it each week. Watch how it changes gradually. This slow observing builds a relationship with place and teaches patience.

Early Birds and Changing Behavior

Bird activity increases long before the air warms. You may notice:
- earlier morning calls
- pairs forming or staying close together
- birds exploring nesting sites
- more confident flight patterns on sunny days

These changes happen because birds respond to daylight more strongly than temperature. Light signals the beginning of reproductive cycles, so their behavior becomes more energetic. Observing this shift can add a sense of companionship to your walks or mornings at home.

. . .

Keeping a simple bird notebook or voice memo can deepen the experience. Record the first day you hear a particular song or see a pair forming. This creates a personal archive of seasonal change.

Subtle Color Changes

Winter colors tend to be muted, but late winter carries hints of brightness. Look for:

- pale green tints on willow branches
- soft red or purple tones on new twigs
- golden light during longer sunsets
- muted greens emerging from moss and grass

These changes occur slowly but consistently.

Color often shifts before warmth arrives.

Practicing awareness of these shades can help the transition from winter's heaviness to spring's movement feel calmer and more hopeful.

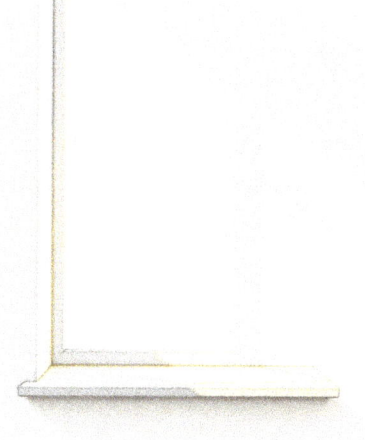

Photography can support this practice. Taking a weekly picture of the same place allows you to see seasonal change more clearly over time.

What seems static in daily life becomes noticeable when compared week by week.

A Modern Observation Practice

You do not need to track every detail. Instead, choose one or two signs of early spring that feel meaningful to you.

> *A single bud, one bird song, a shift in morning light. Let these small moments guide you through the quiet transition from winter into spring.*

You might try:
- a short daily walk focused on color
- a weekly note about something new you saw or heard
- pausing at a window to notice how light falls
- choosing one tree to observe through the season

These practices build a sense of relationship with the natural world. They remind you that renewal happens gradually. You do not need to push yourself into spring. You can arrive with it.

SEASONAL CHECKLIST

First Signs of Spring

A modern, simple checklist readers can screenshot or revisit weekly. Designed to support mindful observation rather than pressure.

Light and Atmosphere
- ☐ Sunrise feels earlier
- ☐ Sunlight has a warmer tone
- ☐ Evenings extend slightly
- ☐ Shadows appear softer
- ☐ Air feels clearer after midday sun

Trees and Plants
- ☐ Buds look rounder or slightly swollen
- ☐ Willow tips shift toward pale green
- ☐ Tiny shoots push through soil or mulch
- ☐ Moss appears brighter
- ☐ First hardy flowers appear (snowdrops, crocus, hellebore)

Birds and Wildlife
- ☐ Birds sing earlier in the morning
- ☐ Pairs begin forming or staying close
- ☐ Birds investigate nesting sites
- ☐ Squirrels or small mammals become more active
- ☐ More birds seen on sunny days, even if the air is cold

Weather and Ground
- ☐ Ice melts in thin sheets rather than thick layers
- ☐ Soil softens slightly during the afternoon
- ☐ Freeze thaw cycles become more noticeable
- ☐ Occasional earthy scents appear on warmer days
- ☐ Puddles form and refreeze overnight

Color Shifts
- ☐ Hints of green along hedges or stream edges
- ☐ Red or purple tones visible on new twigs
- ☐ Golden evening light becomes more common
- ☐ Pale green emerges in grassy patches
- ☐ The landscape feels less monochrome

Your Personal Noticing
- ☐ A moment this week felt lighter or more hopeful
- ☐ You observed a plant or tree more closely than usual
- ☐ You heard a new sound you had not noticed recently
- ☐ You felt a subtle shift in your own energy
- ☐ You sensed the beginning of a new inner season

22 / WORKING WITH FEBRUARY PLANTS AND MATERIALS

FEBRUARY's natural world may appear sparse, but it offers small treasures for creativity, care and connection. Twigs, fallen cones, evergreen sprigs and early buds carry the quiet beauty of late winter. These materials invite simple, hands on practices that help us slow down and notice the season more deeply.

You do not need extensive knowledge of plants or crafting to work with February's offerings.

The goal is not perfection. It is presence. It is the feeling of meeting the outdoors as it is now, without waiting for warmth or flowers. The materials of late winter remind us that even in cold, modest places, there is something to explore.

This chapter offers a handful of small crafts, gentle herbal observations and nature journaling prompts that fit easily into modern life.

Tiny Crafts with Simple Materials

Crafting during February works best when it is light, slow and uncomplicated. The season encourages gentle creativity rather than elaborate projects.

Twig Shapes

Collect a few straight twigs on a walk. Break them or trim them into equal lengths. Bind two or three together with thread or a thin ribbon to form simple shapes such as triangles or small stars. Hang them in a window where they catch returning light.

Evergreen Bundles

If you find fallen pine, juniper or fir sprigs, gather a few into a tiny bundle. Tie them with a small piece of string. Place the bundle near a door or workspace as a reminder of the season's endurance and steady greenery.

Mini Nature Altar

Choose three objects from outdoors that feel meaningful. A stone, a bud, a cone or a strip of bark all work well. Arrange them together on a tray or small dish. This becomes a quiet space of seasonal focus that you can update weekly.

Frozen Light Catchers

If temperatures drop, fill a shallow dish with water, add a few small natural items such as twigs or leaves, and freeze it outdoors. Lift it out once frozen and hold it up to the light. When it melts, return the pieces to nature.

These crafts are not about productivity. They are invitations to slow attention. They help transform winter's simplicity into something gently beautiful.

Herbal Observations for Late Winter

Herbal practices at this time of year rely less on harvesting and more on noticing. Many plants are waking internally long before they show visible signs. Observation becomes a form of partnership with nature.

Buds and Branches

Look for swelling buds on shrubs and trees. Use your fingers to feel their texture. Some may be smooth and tight, others fluffy or scaled. Different species have distinct bud signatures that become easy to recognize over time.

Evergreen Scents

Crush a small fallen evergreen needle between your fingers. Notice how its scent changes in cold air. Some become sharper. Others soften. These scents reflect natural compounds that protect the plant during winter.

Early Greens

In sheltered places, you may spot the earliest shoots of herbs such as chickweed or the very first blades of new grass. They signal that the soil is warming slightly and that microbial life is becoming more active.

Maple or Birch Sap

If you live in a region with sap producing trees, look for signs of early sap flow. A small bead of clear sap on the bark or a glistening patch can hint at the internal movement happening within the tree.

Observing these qualities builds herbal awareness long before harvest season begins.

Nature Journaling Prompts

Nature journaling in February is less about drawing full landscapes and more about collecting impressions. The prompts below work for sketchers, writers or anyone who prefers simple notes.

Prompt 1: The First Thing You Notice

Sit outdoors or by a window for two minutes. Write or sketch the first detail that captures your attention. It may be a sound, a temperature shift, a color or a single branch.

Prompt 2: A Small Change from Last Week

Choose a plant or place you see often. Look for a change that seems too small to matter. Describe it in a few sentences or a quick sketch.

Prompt 3: Late Winter Colors

Make a small palette of the colors you see outdoors now. These may include pale greens, muted purples, soft browns or cold blues. Matching colors deepens seasonal perception.

Prompt 4: What Feels Alive

List or draw three things that feel alive even in the cold. This could be a bird call, a bud, a patch of moss, or even your own breath.

Prompt 5: Messages from the Materials

Hold a twig, leaf or stone from your walk. What feeling or memory does it bring up? Write a short reflection without editing.

Bringing It All Together

February plants and materials are humble yet meaningful. Working with them encourages a slower pace and a deeper sense of belonging to the natural world. You do not need lush greenery or warm days to feel connected.

You only need curiosity and a few minutes of mindful attention.

These tiny crafts, herbal observations and journaling prompts form a simple seasonal practice. They invite you to notice the world as it is now and to prepare gently for the growth that lies ahead.

23 / A PERSONAL FEBRUARY
REFLECTION PAGES

February is a month that invites noticing, gathering and holding small experiences with care.

Its quiet pace encourages reflection without urgency.

It offers a rare space in the year where you can observe who you are in a season of transition rather than accomplishment. Creating a personal record of February is a way to honor your inner landscape alongside the natural one.

Reflection does not have to follow strict structure. It can be a list, a sketch, a memory, a feeling or a small moment that stayed with you. These pages do not ask for perfection. They ask for presence. They help you see how the month has shaped you and how you have moved through it.

What follows is a collection of prompts and creative suggestions that help you build a personal February archive. You can use them as journal entries, notes in a planner, small drawings, voice memos or digital scraps. Choose what feels natural.

Seasonal Memory Keeping

Memory keeping for February often begins with noticing what felt meaningful in its simplicity.

Moments of Light

Recall the mornings or evenings when you first noticed the returning light.

- Did the sunfall look different?
- Was there a moment that felt brighter or softer than usual?

Capture it with a photo, a line of writing or a quick sketch.

Small Natural Encounters

Think of a plant, bird, or natural detail that surprised you.

- A bud rounding
- A bird calling earlier than expected
- A patch of soil softening (*Record one of these encounters in a short memory note.*)

A Shift in Yourself

February often brings internal changes that are subtle but real.

- Did your energy shift?
- Did you crave something new?
- Did you release something old?

These inner changes matter as much as external seasonal signs.

. . .

Creative Entries for the Month

Use these creative ideas to add texture and depth to your February reflections.

Color Strip of the Month

Create a small line of color swatches based on what you saw outdoors.
Three to five shades are enough.
This becomes a visual memory that grows more meaningful over time.

Found Object Collection

Gather a few natural items from the month: a twig, a stone, a fallen cone or a patch of lichen.
Take a photo of them or tape a sketch of them into your journal.
Label each one with a word that describes how the month felt.

Daily One Line

Write one sentence each day about something you noticed.
Not a full journal entry, just a single line.
This builds a simple record that reveals patterns when looked at later.

Voice of the Month

Record a short voice note of any sound that captures the season.
Birdsong, wind, footsteps on frost or the scrape of branches.
Sound can preserve memory in surprising ways.

February Collage

Print a few small photos, clip color samples or sketch tiny shapes.

Arrange them in a loose collage that represents the mood of the month.

This modern approach replaces old scrapbooks with something lighter and more intuitive.

End of Month Reflection Prompts

These prompts help summarize the emotional and sensory shape of February.

1. What felt alive to you this month?
A place, a moment or a sensation.

2. What changed, even slightly?
Inside you or around you.

. . .

3. Which details returned your attention to the present?
   Light, texture, sound or color.

4. What small practice supported you?
   A candle moment, a walk, a pause, a breath.

5. What will you carry forward into March?
   A hope, an intention or a quiet strength.

A Personal February

By the time February ends, you may find that your month is made up not of large events, but of gentle shifts. A personal February is built from noticing, feeling and remembering. It shapes an inner record that supports resilience, clarity and trust in slow beginnings.

Use these pages however you wish.

Add your own prompts. Draw your own lines. This chapter is not an instruction set. It is an invitation to create a small personal archive of a month that often goes overlooked. Through reflection, February becomes a season of its own, one that leaves a subtle but lasting imprint on your year.

*PART 5
APPENDICES*

24 / FEBRUARY MOON PHASES AND SKY NOTES

FEBRUARY'S SKY carries a blend of stillness and subtle motion. The month often feels quiet on the surface, yet the heavens reveal steady seasonal progress. Observing the moon phases and a few key sky patterns can help deepen your connection to this transitional time.

February Moon Phases

The exact dates shift each year, but February often includes:

New Moon

A dark sky that brings clarity and a sense of reset.

The new moon in February encourages reflection and intention setting. It is a quiet moment that aligns naturally with the contemplative tone of the season.

Waxing Crescent

A thin arc of light appears just above the early evening horizon.

This phase often feels hopeful. The light grows gently, mirroring the return of daylight throughout the month.

First Quarter

The moon forms a half circle in the sky.

This is a useful phase for noticing momentum. Energy begins to feel more outward and active as the season shifts toward early spring readiness.

Waxing Gibbous

The moon brightens and enlarges each night.

Its glow reflects increasing momentum in nature. Buds swell more visibly and birds increase their activity during this period of growing light.

Full Moon

Traditionally known as the Hunger Moon.

This is one of February's most striking sky events, illuminating landscapes with sharp, cold clarity. The full moon can make snow glisten and enhance nighttime visibility. It has long been associated with endurance and the final stretch of winter.

Waning Gibbous

The full moon begins to soften and shrink.

This phase supports reflection and release, inviting you to let go of anything that feels heavy or stagnant as the month nears its turning point.

Last Quarter

The moon returns to a half circle.

This is a balancing moment in the lunar cycle, often felt as a time to simplify routines or clear small tasks.

Waning Crescent

A delicate sliver before the next new moon.

This closing phase mirrors February's quiet nature. It encourages rest, gentleness and slow transition toward spring.

February Sky Notes

Long Nights, Clear Views

Cold air creates crisp, steady conditions for stargazing. Constellations appear sharp and luminous, especially Orion, Taurus, the Pleiades and Canis Major.

Shifting Sunrise and Sunset

Light returns noticeably. Sunrise becomes earlier and the evening sky holds color longer. This gradual change influences plant and animal behavior well before warmth arrives.

Quality of Light

Late winter light has a clean, pale tone. Morning skies often appear icy blue, while evenings can hold a soft gold or rose tint.

Visible Planet Activity

Depending on the year, February may offer views of bright planets in the pre dawn or early evening sky. Venus or Jupiter often create striking visual anchors when visible.

Weather and Atmosphere

High, thin clouds sometimes form halos around the moon, signaling incoming moisture. Clear nights are common, allowing frost to settle and enhancing star visibility.

How to Use These Notes

You can track these phases and sky patterns in a journal, check them weekly for insight or simply use them as a gentle guide for observing the season. Each phase and shift represents a small point of connection between your daily life and the natural world.

25 / A GLOBAL LIST OF FESTIVALS AND PLANT MARKERS

FEBRUARY IS a transitional month across many cultures. Although climates and traditions differ widely, certain patterns repeat around the world. Light begins to return in the northern hemisphere, early plant activity stirs beneath the soil and communities mark this turning of the year with celebrations of renewal, cleansing or awakening.

Below is a global overview that pairs February festivals with the plants, materials or natural signs most often associated with them. This map does not attempt to be exhaustive. Instead, it highlights shared themes that appear across continents, reminding readers that the month holds universal patterns of emergence and preparation.

Ireland and Scotland
Brigid's Day and Early Spring Signs**
Festival themes: creativity, light, protection, renewal
Plant markers: rushes, snowdrops, emerging grasses
Seasonal signs: increasing daylight, first lambing, buds beginning to swell

Brigid's Day marks the quickening of the year. Communities traditionally wove crosses or figures from rushes, which still appear in wetland areas even in cold weather.

. . .

Japan

Setsubun and Seasonal Boundaries**

Festival themes: purification, protection, throwing out misfortune

Plant markers: roasted soybeans, holly sprigs, early plum buds

Seasonal signs: subtle warm spells, shifts in wind direction, longer twilight

Setsubun falls near the beginning of spring in the old Japanese calendar. Holly sprigs with beans or thorns once guarded doorways from harm.

China and Diaspora Communities

Lunar New Year (when falling in February)**

Festival themes: renewal, luck, cleansing, family unity

Plant markers: plum blossoms, narcissus, citrus, bamboo

Seasonal signs: early blossoms in warmer regions, fresh fragrances in markets, mild winter rain

Plum blossoms are among the earliest flowers to appear, symbolizing endurance and beauty emerging before warmth arrives.

North America

Groundhog Day and Animal Weather Wisdom**

Festival themes: forecasting, hope for early spring

Plant markers: burrow plants such as clover or roots uncovered by thaw

Seasonal signs: animal movement in warmer spells, freeze thaw cycles, bright winter skies

Although symbolic, this date aligns with mid winter shifts in light that influence wildlife behavior.

Northern Europe

Candle Rituals and Early Thaw Traditions**

Festival themes: light returning, blessing the household, preparing for sowing

Plant markers: willow twigs, hazel catkins, evergreen sprigs

Seasonal signs: first catkins appearing, icicles melting earlier, longer afternoons

Willow and hazel often show early signs of activity, making them key markers of late winter.

Mediterranean Regions

Carnival Season**

Festival themes: feasting, joy, turning winter upside down

Plant markers: grains, spices, citrus, olive branches

Seasonal signs: mild breezes, fresh herb growth in warm pockets, increased bird activity

Carnival traditions grew from a blend of agricultural cycles and winter's final celebrations before spring planting.

India

Late Winter Festivals in February (varies by region)**

Festival themes: learning, color, love, seasonal transition

Plant markers: mustard flowers, blossoming mango buds, early spring herbs

Seasonal signs: warming days, vibrant yellow fields, fragrant breezes

Festivals honoring knowledge, creativity or seasonal color often align with the striking yellow of mustard blossoms.

Indigenous Arctic and Subarctic Communities

Late Winter Observances**

Festival themes: survival, respect for animals, light returning

Plant markers: none in this season, focus instead on snow, ice, and animal behavior

Seasonal signs: sun returning to the sky, shifting sea ice, increased activity among birds and seals

The environment itself is the marker. Light and animal movement shape the seasonal understanding.

. . .

Andean Highlands

Early Seasonal Observances**
Festival themes: honoring water, preparing for agricultural cycles
Plant markers: early shoots of native tubers, high altitude grasses
Seasonal signs: shifts in rain patterns, clearer skies, brightening sun at high elevations

Even in varied climates, the slow shift of light influences mountain communities.

Australia and Aotearoa

Late Summer Observations (southern hemisphere)**
Festival themes: harvest, reflection, gratitude
Plant markers: eucalyptus fragrance peaks, seed pods form, late summer blooms
Seasonal signs: warm winds, long daylight, ripening fruit

February here marks late summer rather than late winter. Plant markers focus on maturity and preparation for the cooler months ahead.

How to Use This Global List

You may find it helpful to notice common themes across regions.
- light returning
- purification or cleansing
- early plant activity
- animal behavior shifting
- celebrations of hope or resilience

These patterns create a sense of belonging to a wider seasonal story. Whether you live in a cold or warm climate, February offers a moment to pause, observe and connect with the natural world in ways both global and deeply personal.

26 / GLOSSARY OF BOTANICAL AND FOLKLORIC TERMS

AERIAL PARTS

The portions of a plant that grow above the soil, including stems, leaves and flowers.

Bud break

The moment when a dormant bud begins to open, signaling active spring growth.

Catkin

A slim, pendulous cluster of flowers, often found on willow, hazel or birch in late winter.

Dormancy

A period of reduced activity in plants, animals or seeds during which growth slows to conserve energy.

Evergreen

A plant that retains its leaves throughout the year, staying green even in winter.

Folklore

Traditional stories, beliefs and observations passed through communities to explain natural events or seasonal change.

Gibbous

A lunar phase where the moon is more than half illuminated but not yet full.

Micro season

A small time period within a larger season, marked by subtle ecological or cultural changes.

Mycorrhizae

The mutually supportive relationship between plant roots and soil fungi that helps transfer water and nutrients.

Perennial

A plant that lives for more than two years, often returning from the same root system each spring.

Photoperiod

The amount of daylight a location receives. Many plants and animals use this cue to regulate seasonal behavior.

Sap flow

The movement of water, minerals and sugars through a plant's vascular system, often increasing in late winter.

Scarcity season

A late winter period when stored food or natural resources are low, referenced in both ecological science and folklore.

Shoot

A young stem or leaf emerging from soil or from a plant's base, indicating active growth.

Solar term

A traditional East Asian time division based on the sun's position, often marking seasonal shifts such as early spring.

Threshold time

A transition period between seasons, often associated with symbolic renewal and subtle natural change.

www.ingramcontent.com/pod-product-compliance
Lightning Source LLC
Chambersburg PA
CBHW081158020426
42333CB00020B/2542